SpringerBriefs in Electrical and Computer Engineering

Rao Mikkilineni

Designing a New Class of Distributed Systems

Springer

Rao Mikkilineni
Kawa Objects Inc.
Los Altos, CA
USA
e-mail: rao@kawaobjects.com

ISSN 2191-8112   e-ISSN 2191-8120
ISBN 978-1-4614-1923-5   e-ISBN 978-1-4614-1924-2
DOI 10.1007/978-1-4614-1924-2
Springer New York Dordrecht Heidelberg London

Library of Congress Control Number: 2011940831

Printed on acid-free paper

Springer is part of Springer Science+Business Media (www.springer.com)

Preface

A new computing model, extends the von Neumann stored program control (SPC) computing model to program and execute self-configuring, self-monitoring, self-healing, self-protecting and self-optimizing (in short, self-managing or self-*) distributed software systems. As opposed to self-organizing systems that evolve based on probabilistic considerations, this approach focuses on the encapsulation, replication, and execution of distributed and managed (regulated) tasks that are specified precisely. Its innately parallel architecture (non-von Neumann), and its architectural resiliency of cellular organisms, are ideally suited to exploit the many-core architecture and low-latency bandwidth networks emerging in the new generation of data centers to improve the price/performance of IT infrastructure by orders of magnitude.

The new computing model (known as the Distributed intelligent managed element (DIME) Network Architecture) consists of a signaling network overlay over the computing service network and allows parallelism between the control (setup, monitoring, analysis and reconfiguration based on workload variations, business priorities and latency constraints) and the computing functions of the distributed software components. A workflow is implemented as a set of tasks, arranged or organized in a directed acyclic graph (DAG) and executed by a managed network of DIMEs. These tasks, depending on user requirements are programmed and executed as loadable modules in each DIME. The distributed software components along with associated profiles defining their use and management constraints are executed by DIMEs endowed with self-management and signaling-enabled-control architecture. The profiles are used as blueprints to setup, execute and control the down-stream DAG at each node based on global and local policies which depend on business priorities, workload fluctuations and latency constraints. A new class of distributed systems with the architectural resilience of cellular organisms are possible using the DIME network architecture with signaling to monitor and regulate the execution of computational workflows with self-* properties.

We describe two proofs-of-concept implementations; one using a new native operating system called Parallax which is specially designed using the DIME computing model to monitor and control multi-core and many-core processors in an Intel processor based server; another using a conventional operating system (Linux) to encapsulate a process as a DIME. The results suggest that the service management can be decoupled from the underlying hardware infrastructure management by utilizing signaling based dynamic configuration of DIME network architecture, and potentially reduce the layers of management software in developing next generation highly-scalable, parallel, distributed virtual service creation, delivery, and assurance platforms.

The purpose of this research brief is to introduce some new ideas based on the study of cellular organisms, human organizational networks and telecommunication networks to improve the resiliency, efficiency and scaling of distributed systems executing various computational workflows. The ideas extend the current von-Neumann computing model by separating the services management from services execution exploiting the parallelism and performance offered by the new class of many-core processors.

The DIME network architecture is a departure from conventional wisdom currently being pursued by the universities and corporate research & development. It adds monitoring and control to each Turing computing node and a parallel signaling enabled network to implement the management of temporal behavior of workflows executed as directed acyclic graphs using a network of managed Turing machines.

The concept of a parallel signaling channel is foreign to the current generation of IT professionals, except for those with telecommunications or voice over IP experience, who are by now either retired or dead. Signaling allows establishing equilibrium patterns and monitor and control exceptions system-wide. It allows contention resolution based on system-wide view and eliminates race conditions and other common issues found in current distributed computing practice. In systems with strong dynamic coupling between various elements of the system, where each change in one element continually influences other element's direction of change, signaling in the computational model helps implement system-wide coordination and control based on system-wide priorities, workload fluctuations and latency constraints.

We have demonstrated the feasibility of this approach using two prototypes. However, in order to take these concepts to practical application in mission critical environments, the DIME network architecture requires its validation and acceptance by a larger community. This research brief presents the concept and the results for such validation and analysis.

Acknowledgments

I am extremely indebted to Dr. Giovanni Morana, Daniel Zito and Marco Di Sano from the University of Catania, for their contributions to the development of the DIME Network Architecture and the prototype in Linux operating system. I am also grateful to Ian Seyler from Return Infinity who developed the native operating system from scratch implementing the DIME network architecture and demonstrated self-repair, auto-scaling and live migration in multi-core servers without the use of Hypervisors. All the prototype work was carried out by individuals in their spare time, interested in exploring new ideas that go beyond conventional thinking in spite of the risk involved, with no funding from any organization. I also wish to acknowledge many valuable discussions with and encouragement from Kumar Malavalli, and Albert Comparini from Kawa Objects Inc.

Acknowledgments

About the Author

Dr. Rao Mikkilineni received his PhD from University of California, San Diego in 1972 working under the guidance of prof. Walter Kohn (Nobel Laureate, Chemistry, 1998). He later worked as a research associate at the University of Paris, Orsay, Courant Institute of Mathematical Sciences, New York and Columbia University, New York. He is currently the Founder and CTO of Kawa Objects Inc., California, a Silicon Valley startup developing next generation computing infrastructure. His past experience includes working at AT&T Bell Labs, Bellcore, U S West, several startups and more recently at Hitachi Data Systems.

Dr. Mikkilineni co-chairs the 2nd "Convergence of Distributed Clouds, Grids and their Management" conference track in IEEE International WETICE 2012 Conference. Current work on DIME network architecture was first presented in WETICE 2010 in Larissa, Greece based on the workshop discussions started in WETICE 2009 in Groningen, The Netherlands.

Contents

Chapter 1
An Introduction to the Design of a New Class of Distributed Systems

Abstract Globalization, communication, collaboration, and commerce at the speed of light are creating a demand for highly resilient, efficient, and scalable distributed transactions through various applications such as high frequency trading, social networking, and federated enterprise business process automation. Current computing models based on serial von-Neumann stored program control implement the services and their management in a serial fashion. On the other hand, the dynamic coupling between various elements of the system, where each change in one element continually influences some other element's direction of change, introduces a highly temporal element and parallelism to the distributed transactions. This research brief introduces a new distributed computing model that exploits the parallelism and performance offered by the new generation of many-core processors to improve the resiliency, efficiency, and scaling of distributed transactions.

Communication, Collaboration, and Commerce at the Speed of Light

"There is nothing more difficult to take in hand, more perilous to conduct, or more uncertain in its success, than to take the lead in the introduction of a new order of things."—Niccolo Machiavelli, "The Prince" 1469–1527

High speed computers connected with low latency networks today, can execute 1,000 trades per second; exchanges can process trades in less than 500 μs (or millionths of a second). By 2010 High Frequency Trading accounted for over 70% of equity trades taking place in the US. In high-frequency trading, programs analyze market data from disparate sources in real-time to capture trading

R. Mikkilineni, *Designing a New Class of Distributed Systems*, SpringerBriefs
in Electrical and Computer Engineering, DOI: 10.1007/978-1-4614-1924-2_1,
© The Author(s) 2011

opportunities that may open up for only a fraction of a second to several hours. In essence, the high-frequency traders utilize distributed computing networks of stored program computing elements whose resources are shared to execute business processes to accomplish a common goal. Other examples of distributed computing are the social networking applications such as Facebook, Twitter etc., which connect millions of geographically separated end users to communicate and collaborate transcending national and cultural boundaries. Enterprises use federated systems to collaborate and execute common business processes spanning across distributed resources belonging to different owners.

Global connectivity, resulting communication, collaboration, and commerce at the speed of light have created the "network effect" which impacts the result of a transaction between participants much faster than the increase of the number of participants. Originally the network effect was used by Theodore Veil [1] to gain monopoly in telephone service and it was later popularized by Robert Metcalfe [2] with his law which states that the cost of the network increases with the number of participants as N while the value of the network increases as N^2. Later, Rod Beckstrom presented a mathematical model [3] for describing networks that are in a state of positive network effect and also the "inverse network effect'" with an economic model. "Beckstrom's Law" says that the value of a network is the net value of each user's transaction summed up for all users. At its core, the concept is about transactions: The value for users is the total benefits from all transactions in a network minus the cost of those transactions. The network effect is a temporal many body effect with complex interactions and the result depends on not only the nature of distributed transactions but also on how they are managed.

Sharing of resources and collaboration through distributed transactions, while they provide leverage and synergy, also pose problems such as contention for same resources, issues of trust, and management of the impact of latency in communication among the participants. The discipline of distributed computing systems addresses these issues.

Distributed Systems and their Management

Distributed systems, in essence, consist of a set of physically and may be geographically distributed autonomous components that communicate and collaborate to accomplish a goal using local and/or remote and private and/or shared resources in an optimal fashion. Distributed systems consist of consumers and suppliers exchanging services using local and/or remote resources and can compose them to create other value added services. The resulting shared transactions among a network of consumers and suppliers require an orderly managed process for communication, collaboration, contention resolution, and assurance of end to end transaction integrity that addresses each transaction's fault, configuration, accounting, performance and security (FCAPS) constraints defined based on the nature of the transaction. The end-to-end transaction-level integrity mandates a stringent system-wide distributed management infrastructure in addition to component level FCAPS management and service execution.

Distributed systems support both component level autonomy to optimally utilize the resources and system-wide collaboration and coordination to accomplish the system-level goal through distributed transactions. The autonomy assumes that each entity participating as an actor in a distributed system has certain abilities such as control over local resources to provide the service, communication capability to participate in the transaction, and the ability to define and execute the services. The distributed transaction is defined by the end-to-end transaction goal that utilizes the component resources, and the nature of the transaction defines the end-to-end resource coordination and management. There are transactions where perception, cognition and action are clearly separated. There are other transactions in which there is dynamic coupling between various elements of the system, where each change in one element continually influences every other element's direction of change. These transactions tightly integrate a system's sensory and control functions with analysis. High frequency trading is one such example where monitoring and control of various components in the distributed system are performed in microseconds. In addition, when the distributed system supports a large number of transactions, additional management of resources to resolve contention based on end-to-end goal requirements, latency constraints and environmental changes, become essential. The highly temporal nature of distributed computing, with dynamic coupling, connectivity and system-wide coordination introduces a natural element of parallelism between a service transaction and its management.

In summary, distributed systems are characterized by a set of autonomic loosely coupled self-managing nodes executing concurrent tasks collaborating and coordinating to support highly temporal distributed transactions accomplishing a system-wide goal. The resiliency of the distributed system depends heavily on both the node resiliency and the network resiliency. Current computing models based on serial von-Neumann stored program control implement the services and their management in a serial fashion. For example, an operating system's task for resource allocation such as open(), and close () are mixed with service execution tasks such as read() and write(). All these instructions are serially processed by the von Neumann computing model implementing a Turing machine. This approach gives raise to many of the current distributed computing practices. However there are two major drivers forcing a reexamination of how distributed systems are managed:

1. According to Prahlad and Krishnan [4], in an enterprise that is competing globally, the traditional sources of advantage—access to technology, labor and capital—are no longer unique differentiators for most firms. They say that the "new source of competitive differentiation may lie in the internal capacity to reconfigure resources in real-time." This mandates the real-time management of business process execution which in turn demands 100% availability, reliability, performance optimization and security of the distributed computing infrastructure implementing the business processes.

2. Communication, collaboration and commerce at the speed of light are driving the need for massively scalable service deployment, and the increasing demand to do "more with less" is driving the infrastructure consolidation and new approaches to distributed computing efficiency.

The demand for improving distributed transaction resiliency (of FCAPS management), efficiency, and scaling are pushing for a search for new approaches to distributed computing practice.

Other distributed systems in nature such as cellular organisms, human organizational networks and telecommunication networks provide clues to designing highly resilient, efficient and scalable distributed computing systems. This research brief examines these distributed systems and their management to identify key abstractions and patterns that contribute to their resiliency, efficiency, and scaling. All the three examples presented have a set of common abstractions that are essential in exhibiting the common characteristics of high resiliency, efficiency, and scaling. These attributes are:

1. A parallel overlay of a management channel to facilitate system-wide collaboration, coordination and control to assure end-to-end transaction integrity and successful completion of system-wide goals.
2. Both component level autonomy to optimally manage local resources and system-wide management to optimally share the local and remote resources to accomplish system-wide goals.

We exploit these abstractions to design a new class of distributed systems offering the architectural resiliency of cellular organisms, the scalability of human organizational networks and the telecom grade trust and efficiency of telecommunication networks. We propose a novel and interesting non-von Neumann computing model that exploits parallelism of the new generation multi-core and many-core processors with high bandwidth inter-process communications. The new computing model separates service management from the service execution and exploits an overlay of signaling network to provide system-wide collaboration, coordination and control. This approach extends the current serial von-Neumann computing model by adding the management overlay and thus avoids making changes to the current service execution paradigms.

This approach is quite distinct from current grid and cloud computing approaches [5, 6] which implement automation of services management using a mixture of node level service management (which uses serial von Neumann computing model) and a plethora of resource management systems to coordinate and control network level service management. We argue that while these approaches are successful to a certain extent, they suffer from limitations of resiliency, efficiency, and scaling. We show that the separation of services management from their execution allows a decoupling of services management and control from the underlying hardware infrastructure management and control. Thus the new computing model allows the design and execution of resilient services using not so reliable hardware infrastructure just as the cellular organisms do. This has profound implications on next generation hardware design which could eliminate the need for special purpose hardware, hardware clustering and extensive resource management strategies for providing FCAPS resilience [7].

We demonstrate the feasibility of this new approach by implementing auto-scaling, self-repair and self-management features in two cases:

1. By implementing the new computing model in Linux operating system to encapsulate a process demonstrating dynamic reconfiguration of services and their management and
2. By implementing the new computing model in a native operating system called Parallax in a multi-core Intel server written in assembler language with C and C++ interfaces also demonstrating dynamic reconfiguration of services and their management.

These two prototypes demonstrate the feasibility of the new approach. However, in order to take these concepts to practical application in mission critical environments, the DIME network architecture requires its validation and acceptance by a larger community. This research brief presents the concept and the results for such validation and analysis.

Evolution, Revolution and the Adoption of Disruptive Innovation

The IT landscape is filled with unfulfilled technology promises, surprise winners and the meteoric rise and sudden fall of various technology companies. Evolution has a way of selecting best practices that lower system's entropy and assure its survival. Sometimes it prefers incremental improvements to optimize in its current local minimum. Other times it prefers a new local minimum that is radically different from its current equilibrium.

Technology or process innovation that improves productivity (and lowers the entropy) evolves through three distinct phases (namely, the incubating, emerging and mature phases) which have different returns on investment. Disruptive technologies that raise the productivity from one level to a next higher level occur through evolutionary need for competitiveness. Figure 1.1 summarizes the three phases of innovation in Information Technologies.

Traditionally, as technologies start to mature, governments and corporations have devoted a part of their revenues (taxes or profits) in incubating technologies as an investment to their future competitiveness and survival. History has shown that this investment is about 3–6% of their revenue. History also has shown that such investment attracts the creative scientists and engineers to nurture the culture of collaboration and structures that cultivate talent (without the near-term profit oriented cut-throat competition) and go on to achieve Nobel prizes and National Science Awards. Corporate R&D, University research, DARPA and NSF funded projects, traditionally fulfilled this role.

Fig. 1.1 Technology innovation and adoption phases

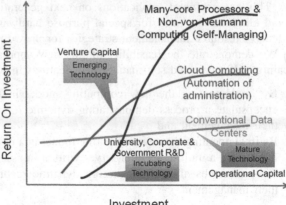

The figure shows three different stages of technology (S-curve) evolution. The conventional data centers using various server, network and storage resource management strategies is considered mature technology since the productivity improvements have reached a plateau. The cloud computing technology is emerging to provide productivity gains through automation of various service management tasks going beyond resource management. The many-core processors and the new opportunity to exploit the resulting hardware advances with parallelism and performance offer a new opportunity that is currently in the incubating stage.

The evolution from one phase to another is gradual. When the incubating technologies start to show promise as emerging technologies, the animal spirits of Venture Capitalists start smelling high profits and exploit entrepreneur's product development expertise and establish product leadership. The culture and structure required in the incubating phase are different (at least it was in the past when institutions such as AT&T Bell Labs invested a portion of their revenue in research for research sake) from the culture and structure required for exploiting emerging technologies.

As the products, processes, services and technologies prove themselves in customer environments, conventional capitalism kicks in and large corporations exploit scale through establishing operational excellence and customer intimacy. Again the evolution emphasizes that the culture and structure that is exploited by conventional capitalism are different from the cultures that support disruptive innovation and establishing product leadership. The culture of control that helps in establishing operational excellence fails miserably in creating disruptive innovation.

The long and short of the discussion is that the patterns of evolutionary advantage are different in supporting different phases of implementing productivity improvements and establishing competitive differentiation. We believe that exploiting the hardware revolution with many-core processors with the non-von

Neumann computing model proposed here is part of disruptive innovation that can provide orders of magnitude improvement in resiliency, efficiency, and scaling of distributed computing systems. However, a lack of immediate profits and the potential risk make it an incubating technology that requires a different approach from projects that have immediate commercial success for its validation and acceptance.

Organization of this Research Brief

The purpose of this research brief is to introduce some new ideas based on the study of cellular organisms, human organizational networks and telecommunication networks to improve the resiliency, efficiency, and scaling of distributed systems executing various computational workflows. The ideas extend the current von-Neumann computing model by separating the services management from services execution exploiting the parallelism and performance offered by the new class of many-core processors.

Current Information Technology solutions have become silos of server, storage and network infrastructure with poor end-to-end distributed transaction reliability, availability, performance and security as recent episodes at Sony, Amazon, Google, and RSA [8–11] demonstrate. The service failures and security breaches caused big losses and major disruptions to the affairs of their customers. We believe that there is a need for reexamining the fundamental architectural foundation of Information Technologies to transform the data centers from their current role of being just managed server, networking, and storage hosting centers (whether physical or virtual), to true service switching centers with telecom grade trust. Current advances in many-core processors and the associated power and space savings offer an opportunity to refresh the current data centers with new innovation. Current operating systems and management paradigms are not adequate to leverage the full potential of the hardware innovation offered by these many-core processors. We need a paradigm shift from resource switching and connection management to services switching and service connection management. We also believe that new approaches are essential to replace the current efforts to replicate the complexity inside the data center today, also inside the many-core servers. The organization of the rest of the research brief is aimed at making a case for a paradigm shift to a new class of distributed systems design and execution.

Chapter 2 provides a review of current state of the art and science of distributed systems and understanding of distributed systems design to make a case for the need to change. Chapter 3 introduces the new non-von Neumann computing model derived from studying the cellular organisms, human organizational networks and telecommunication networks which use signaling and self-management to improve their resiliency, efficiency, and scaling. Chapter 4 provides a discussion of two implementations of the new computing model to demonstrate feasibility. The first implementation demonstrates the non-von Neumann computing model in a conventional operating system (Linux) which allows the migration of current service architectures with minimal disruption. The second approach demonstrates a new native operating system that exploits the parallelism of the many-core servers to create a new class of resilient, efficient and scalable distributed systems. Chapter 5 presents some conclusions based on our experience with the implementations and points to a few new research directions to take these ideas forward.

Only time will tell if these ideas will bear fruit. But again, as Mitchell Waldrop points out, revolutions are not revolutions if they are believed in at the start. Are they?

"How did it go in Berkeley? Did they like your ideas?"
"It was the pits," said Arthur. "Nobody there believes in increasing returns."
Susan Arthur had seen her husband returning from the academic wars before.
"Well," she said, trying to find something comforting to say, "I guess it
Wouldn't be a revolution; would it, if everybody believed in it at the start?"
   —Waldrop, M.M., "complexity: The Emerging Science at the Edge of
   Order and Chaos", New York, Simon and Schuster, (1992) p 19.

References

1. http://www.corp.att.com/history/history1.html
2. http://www.hyperorg.com/backissues/joho-jan19-01.html#reed
3. It's All In Your Head. Forbes. 2007-05-07. http://www.forbes.com/forbes/2007/0507/052.html. Retrieved 2010-12-10
4. C.K. Prahlad, M.S. Krishnan, *The New Age of Innovation* (McGraw Hill, New York, 2008), p. 9
5. R. Buyya, R. Ranjan, Special section: Federated resource management in grid and cloud computing systems. Future Gener. Comput. Syst. **26**, 1189–1191 (2010)
6. R. Buyya, C.S. Yeo, S. Venugopal, J. Broberg, I. Brandic, Cloud computing and emerging IT platforms: Vision, hype, and reality for delivering computing as the 5th utility. Future Gener. Comput. Syst. **25**(6), 599–616 (2009)
7. V. Sarathy, P. Narayan, R. Mikkilineni, "Next Generation Cloud Computing Architecture: Enabling Real-Time Dynamism for Shared Distributed Physical Infrastructure," WETICE, (2010) pp. 48–53, 19th IEEE International Workshops on Enabling Technologies: Infrastructures for Collaborative Enterprises, 2010
8. C. Morris, Sony playstation facing yet another security breach, New York, CNBC.com (2011), http://www.cnbc.com/id/43079509
9. P. Thibodeau, J. Vijayan, Amazon EC2 service outage reinforces cloud doubts. Computerworld (2011), http://www.computerworld.com/s/article/356212/Amazon_Service_Outage_Reinforces_Cloud_Doubts
10. http://www.searchenginejournal.com/googles-downtime-affected-5-ofthe-internet/10463/
11. A. Moscaritolo, RSA confirms Lockheed hack linked to SecurID breach (2011), SC MAGAZINE, June 07, http://www.scmagazineus.com/rsa-confirms-lockheed-hack-linked-to-securidbreach/article/204744/)

Chapter 2
Understanding Distributed Systems and Their Management

Abstract An analysis of current implementation of distributed computing and its management using the von-Neumann stored program control computing model is presented to identify ways to improve the resiliency, efficiency and scaling of distributed transactions supporting the demands of communication, collaboration and commerce at the speed of light. A comparison with other distributed computing models such as the cellular organisms, human organizational networks and telecommunication networks points to a new computing model that leverages parallelism, signaling and end-to-end transaction management improving the resilience, efficiency and scaling of distributed transactions.

Stored Program Control Computing Model, Distributed Computing, and Management

"Despite more than 30 years of progress towards ubiquitous computer connectivity, distributed computing systems have only recently emerged to play a serious role in industry and society. Perhaps this explains why so few distributed systems are reliable in the sense of tolerating failures automatically, or guaranteeing properties such as high availability, or having good performance even under stress, or bounded response time, or offer security against intentional threats. In many ways the engineering discipline of reliable distributed computing is still in its infancy.

Reliability engineering is a bit like alchemy. The field swirls with competing schools of thought. Profound arguments erupt over obscure issues, and there is little consensus on how to proceed even to the extent that we know how to solve many of the hard problems."—Kenneth Paul Birman [1].

"The sharing of resources is a main motivation for constructing distributed systems"— George Coulouris, Jean Dollimore, Tim Kindberghy [2].

R. Mikkilineni, *Designing a New Class of Distributed Systems*, SpringerBriefs 9
in Electrical and Computer Engineering, DOI: 10.1007/978-1-4614-1924-2_2,
© The Author(s) 2011

"A distributed system is a collection of independent computers that appears to the users as a single coherent system."—Andrew S. Tanenbaum and Martin van Steen [3]

Since von Neumann [4] discussed, at the Hixon symposium in 1948, his views on the resilience of the cellular organisms and the shortcomings of the stored program control machines he designed, the quest for building reliable systems on an infrastructure that is not, often, reliable has been the holy grail of information technologies. As Birman points out, distributed systems pose more of a challenge because of the need for the management to coordinate the collaborating resources that span across different hardware and geographies to accomplish the goals of the overall system. A transaction to accomplish a system goal in a distributed system by definition spans across distributed shared resources transcending geographical boundaries. Fault, configuration, accounting (of who uses what resources), performance and security management of distributed resources that collaborate using different communication mechanisms and network connections determine the overall response time and success or failure of the end-to-end transaction. The shortcomings or, the aspects of alchemy as Birman [1] puts it, arise from a lack of consistent treatment of distributed computing and its management. In this chapter we revisit the stored program control computing model, traditional treatment of distributed systems, and their management strategies to show that many of the issues are a result of the coupling of computing tasks and their management tasks implemented using the stored program control computing model. We also study other distributed systems such as cellular organisms, human network organization and telecommunication networks that are proven to be very resilient, to define a new class of distributed computing systems.

Current distributed computing practices have their origin from the server-centric von Neumann Stored Program Control (SPC) architecture that has evolved over the last five decades. In its simplest form, the computation and storage are separated using CPU and memory devices. A single storage structure holds both the set of instructions on how to perform the computation and the data required or generated by the computation. Such machines are known as stored-program computers. The separation of storage from the processing unit is implicit in this model. The ability to treat instructions as data is what makes compilers possible. It is also a feature that can be exploited by computer viruses when they add copies of themselves to existing program code. The problem of unauthorized code replication can be addressed by the use of memory protection support. Virtual memory architectures have incorporated management of computing and storage resources in the operating systems. During last five decades, many layers of computing abstractions have been introduced to map the execution of complex computational workflows to a sequence of 1 s and 0 s that eventually get stored in the memory and operated upon by the CPU to achieve the desired result. These include process definition languages, programming languages, file systems, databases, operating systems etc.

While this has helped in automating many business processes, computing remained mainly centralized in islands of main frames and mini computers with occasional time-sharing thrown in until Ethernet was introduced to connect multiple computers. Distributed computing utilizing networked computing resources came of age in the 1970s, starting from a client–server computing model, and has fully developed to current grid-computing and cloud computing implementations where hundreds of physical and virtual servers are used in distributed grids and clouds to provide orchestrated computational workflow execution. With the steady increase of computing power in each node and connectivity bandwidth among the nodes, during the last three decades, sharing of distributed resources by multiple applications to increase utilization has improved the overall efficiency in implementing business processes and workflow automation. Sharing of resources and collaboration, provide leverage and synergy, but also pose problems such as contention for same resources, failure of participating nodes, issues of trust, and management of latency and performance. These problems are well articulated in literature and the discipline of distributed computing is devoted to address them. There are three major attributes that must be considered in designing distributed systems:

1. Resiliency: Collaboration of distributed shared resources can only be possible with a controlled way to assure both connection and communication during the period of collaboration. In addition, the reliability, availability, accounting, performance, and security of the resources have to be assured so that the users can depend on the service levels they have negotiated for. The FCAPS management allows proper allocation of resources to appropriate consumers consistent with business priorities, requirements and latency constraints. It also assures that the connection maintains the service levels that are negotiated between the consumers and the suppliers of the resources. Resiliency therefore consists of the ability to:

   a. Measure the FCAPS parameters both at the individual resource level and at the system level and
   b. Control the resources system-wide based on the measurements, business priorities, varying workloads, and latency constraints of the distributed transactions.

   In an ideal environment, resources are offered as services and consumers who consume services will be able to choose the right services that meet their requirements or the consumers will specify their requirements and the service providers can tailor their services to meet consumer's requirements. The specification and execution of services must support an open process where services can be discovered and service levels are matched to consumer requirements without depending on the underlying mechanisms in which services are implemented. In addition service composition mechanisms must be available to dynamically create new value added services by the consumers.

2. Efficiency: The effectiveness of the use of resources to accomplish the overall goal of the distributed system depends on two components:

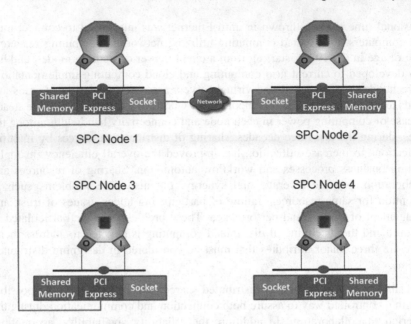

Fig. 2.1 A network of stored program control nodes implementing computational workflow and the resource management workflow

a. Individual resource utilization efficiency which measures the cost of executing a task by the component with a specified service level and

b. The coordination and management cost which assures that the distributed components are contributing to the overall goals of the system with specified end-to-end transaction service level.

The efficiency is measured in terms of return on investment (ROI) and total cost of ownership (TCO) of the distributed system.

3. Scaling: As the requirements in the form of business priorities, workload variations or latency constraints change, the distributed system must be designed to scale accordingly. The scaling may involve dialing-up or dialing-down of resources, geographically migrating them and administratively extending the reach based on policies that support centralized, locally autonomous or a hybrid management with coordinated orchestration

Therefore, strictly speaking, distributed computing, should address, (a) the computational workflow execution implemented by the SPC computing nodes and (b) the management workflow which addresses the management of the network of computing nodes to assure connectivity, availability, reliability, utilization, performance, and security (FCAPS) of the distributed resources. Figure 2.1, shows a group of SPC nodes connected by different mechanisms (shared memory, PCI Express and Socket communications) with different bandwidths.

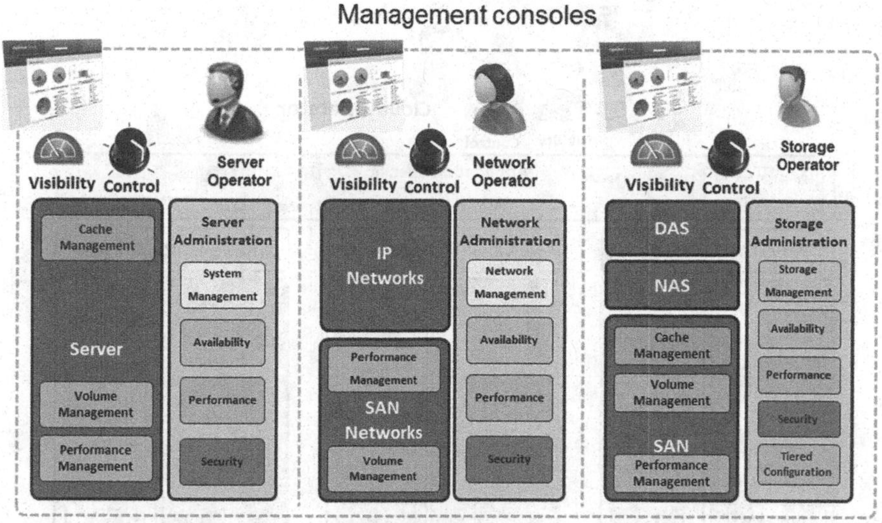

Management consoles

Fig. 2.2 Layers of management governing the behavior of computational workflows with system, server, network and storage management systems

Traditionally, the operating system provided the node resource management (memory, CPU cycles, bandwidth, storage capacity, throughput and its rate of Input/Output). Various network, storage and server management systems provided the FCAPS management of resources at the network or system level. As the distributed transactions start spanning across many nodes and networks across various geographies, the end to end response becomes a function of workloads on individual nodes, sub-networks, business priorities that control access to various applications sharing the resources and the overall latency constraints. In order to control the end-to-end transaction FCAPS, both management workflows and computational workflows are implemented using the same SPC computing network. Many of the functions controlling the provisioning, fault management, utilization control, performance and security monitoring and control are distributed in various nodes as workflows using the same SPC nodes.

The server, network and storage resources are monitored and controlled by different management workflows often duplicating many functions. The open systems approach, multiple vendor products evolving simultaneously to specialize in server, network and storage functions to improve resiliency, efficiency and scaling, and a lack of end-to-end systems optimization strategies in point products, all have contributed to creating a complex web of hardware and software systems in the IT data center. Figure 2.2 shows the evolution of the management workflows.

With the advent of virtualization technologies, the resource management is accessible through automation systems reducing the labor intensive server, network and storage management functions thus reducing the human latency involved

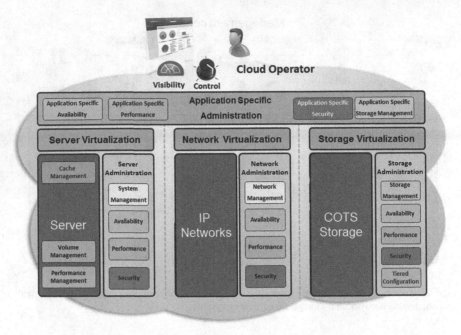

Fig. 2.3 Automation of application-centric management governing the behavior of computational workflows with virtual server, network and storage management systems

in responding to changing circumstances. Both grid and cloud computing technologies make use of system-wide resource management to control the resource allocation priorities based on specific application requirements such as business priorities, workload fluctuations and latency constraints. The management systems monitor the resource utilization characteristics across the system and implement appropriate control workflows to reconfigure the resources. Figure 2.3 shows the evolution of resiliency with the advent of grid and cloud computing.

As the number of CPUs within an enclosure increase, the distributed resource control within the enclosure (processors and cores in each processor) is relegated to the local operating system in that enclosure and it implements appropriate management workflows to match the demands of applications that request the resources.

The system level management is implemented using a plethora of resource management systems. The resulting layers of computational and management workflows improve resiliency, efficiency and scaling of distributed systems. Figure 2.4 shows the evolution of resiliency, efficiency and scaling with the introduction of grid and cloud computing technologies.

The picture shows the FCAPS management ability and the resulting resiliency of a distributed system, its scaling ability to be able to add the number of computing elements and the efficiency (shown by the size of the sphere). Conventional computing where server, network and storage management are resource centric

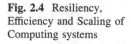

Fig. 2.4 Resiliency,
Efficiency and Scaling of
Computing systems

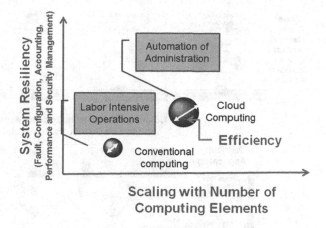

and are static (i.e., applications do not have the ability to request more or less resources based on their workloads and latency constraints), the resiliency and scaling are still limited by the human latency involved in resource management using various management systems. The physical server is the computing unit in conventional computing which provides FCAPS management. With the advent of virtualization technologies, another layer of application-centric dynamic resource management albeit using multiple management systems, improves the resilience, efficiency and scaling. In this case, a virtual server constitutes a computing unit with FCAPS management.

In both cases, the computational workflow and management workflow are implemented using the SPC computing node (either a physical server based application or virtual server based application) and are executed in a serial fashion as shown in Fig. 2.5.

In this approach, the end-to-end distributed transaction resilience is dependent on detecting changes required and correcting them and the serial nature of this process introduces an inherent latency in the process. If the timescale, in which the external requirements change because of the changes in business priorities, workloads fluctuations or latency constraints, is much larger than the time it takes to respond, the serial process does not pose a serious problem. On the other hand, if the scale of changes varies by orders of magnitude in a short period of time as experienced by current Internet based service delivery on a global scale, the von-Neumann bottleneck becomes pronounced and will adversely impact the resiliency, efficiency and scaling of distributed transactions.

The limitations of the SPC computing architecture were clearly on his mind when von Neumann gave his lecture at the Hixon symposium in 1948 in Pasadena, California [4]. He pointed out that "Turing's procedure is too narrow in one respect only. His automata are purely computing machines. Their output is a piece of tape with zeros and ones on it." However, he saw no difficulty in principle in dealing with the broader concept of an automaton whose output is other automata

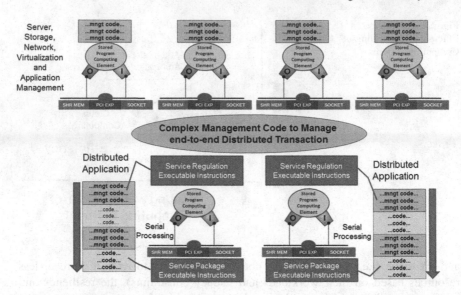

Fig. 2.5 Distributed computational and management workflow implementation using the stored program control computing model where the node is either a physical server or a virtual server

and in deriving from it the equivalent of Turing result. He went on to make this remark. "Normally, a literary description of what an automaton is supposed to do is simpler than the complete diagram of the automaton. It is not true a priori that this always will be so. There is a good deal in formal logic which indicates that when an automaton is not very complicated the description of the function of the automaton is simpler than the description of the automaton itself, as long as the automaton is not very complicated, but when you get to high complications, the actual object is much simpler than the literary description." He went on to say, "It is a theorem of Gödel that the description of an object is one class type higher than the object and is therefore asymptotically infinitely longer to describe." The conjecture of von Neumann leads to the fact that "one cannot construct an automaton which will predict the behavior of any arbitrary automaton [5]." This is the case with the Turing machine implemented by the SPC model.

Cellular Organisms, Genes, Chromosomes and Distributed Computing

It turns out that the description and the execution of the described function play a crucial role in cellular organisms giving them the capability to replicate, repair, recombine and reconfigure themselves. These genetic transactions are supported by DNA (Deoxyribonucleic acid), genes and chromosomes. As Mitchell Waldrop explains in his book on Complexity [6], "the DNA residing in a cell's nucleus was not just a blue-print for the cell—a catalog of how to make this protein or that protein. DNA was actually the foreman in charge of construction. In effect, was a kind of molecular-scale computer that directed how the cell was to build itself and repair itself and interact with the outside world." The conjecture of von Neumann leads to the fact that the SPC computing model alone is not adequate for self-replication and self-repair. Organisms somehow have managed to precisely encapsulate the descriptions of building and running a complex system such as a human being in a simpler vehicle such as a set of genes and chromosomes. They have also managed to invent mechanisms for replication, repair, recombination and rearrangement to execute the descriptions precisely. According to Jacob and Monod [7], "The gene circuitry of an organism connects its gene set (genome) to its patterns of phenotypic expression. The genotype is determined by the information encoded in the DNA sequence, the phenotype is determined by the context dependent expression of the genome, and the circuitry interprets the context and orchestrates the patterns of expression. Gene circuits sense their environmental context and orchestrate the expression of a set of genes to produce appropriate patterns of cellular response."

The relationship between von Neumann's self-replication and genetic behavior was recognized by Chris Langton [8] who created a new field called artificial life, which evolved to throw light on self-organization and the emergence of order from chaos. The field of artificial life and genetic programming focus on self-organization under probabilistic evolution rules that reduce overall entropy of the system. However, equally fascinating feature of the genome is its ability to reproduce itself with a precision that is unparalleled.

As George Dyson, in his book 'Darwin among the Machines,' observes [9] "The analog of software in the living world is not a self-reproducing organism, but a self-replicating molecule of DNA. Self-replication and self-reproduction have often been confused. Biological organisms, even single-celled organisms, do not replicate themselves; they host the replication of genetic sequences that assist in reproducing an approximate likeness of them. For all but the lowest organisms, there is a lengthy, recursive sequence of nested programs to unfold. An elaborate self-extracting process restores entire directories of compressed genetic programs and reconstructs increasingly complicated levels of hardware on which the operating system runs." Life, it seems, is an executable directed acyclic graph (DAG) and a managed one at that.

A cell is formed by a stable pattern of chemical molecules that establish equilibrium with its environment and optimize resource utilization to maintain its equilibrium. According to Richard Dawkins [10], "DNA molecules do two important things. Firstly, they replicate, that is to say they make copies of themselves. This has gone on non-stop ever since the beginning of life and the DNA molecules are now very good at it indeed."

It is one thing to speak of the duplication of DNA. But if the DNA is really a set of plans for building a body, how are the plans put into practice? How are they translated into the fabric of the body? Dawkins poses these questions and answers them. "This brings me to the second important thing DNA does. It indirectly supervises the manufacture of a different kind of molecule—protein. The coded message of the DNA, written in the four-letter nucleotide alphabet, is translated in a simple mechanical way into another alphabet. This is the alphabet of amino acids which spells out protein molecules."

The indirect supervision task mentioned by Dawkins is related to the function of the Gene, which is a piece of DNA material that contains all the information needed to "build" specific biological structure. Genes thus contain the information to make proteins, the body's building blocks. Proteins make up the structure of the organs and tissues; they are also needed for the body's chemical functions and pathways. Each protein performs a specific job in the body's different types of cells, and the information for making at least one protein is contained in a single gene. The pattern or sequence of the genes is like a blueprint that tells the body how to build its different parts. The Genes constitute the workflow components for building the biological system consisting of a group of cells that act with a single purpose which is to propagate the equilibrium patterns they have found to survive.

Cells may have a variety of fates: they may divide and increase in number, differentiate into different kinds of cells, or die (apoptosis). "Determination of the fate of a cell starts when a protein called a signaling molecule binds to a receptor embedded in the cell membrane," says Yasushi Sako, [11] Chief Scientist, Cellular Informatics Laboratory, RIKEN Advanced Science Institute in Japan. When bound by a signaling molecule, the receptor is activated and information is transmitted into the cell. The information is then conveyed from one protein to another within the cell through repeated binding, dissociation and migration until it eventually reaches the cell nucleus, where it induces the expression of a specific gene. This gene triggers various cellular responses, including proliferation, growth inhibition, differentiation, apoptosis, and oncogenic transformation.

Even the simplest unicellular organism provides a good example of self-management and reproduction to sustain life. However, more recent studies in evolutionary developmental biology throw fundamental insights into the inner workings of how groups of cells are organized and orchestrated to create what the biologist Sean B. Carroll calls "endless forms more beautiful" [12]. He points out that "just about 1.5%, codes for the roughly 25,000 proteins in our bodies. So what else is there in the vast amount of our DNA? Around 3% of it, made up of about 100 million individual bits, is regulatory. This DNA determines when, where, and how much of a gene's product is made." With modularity, multi-functionality and redundancy built in its architecture, how does orchestration take place? The orchestration is accomplished by "parallel and sequential actions of tool kit genes—dozens of genes acting at the same time and same place, many more genes acting in different places at the same time, and hundreds of toolkit genes acting in sequence." Each gene may have multiple switches which can be switched on or off by the gene toolkit thus controlling the behavior of the gene. "The developmental steps executed by individual switches and proteins are connected to those of other genes and proteins. Larger sets of interconnected switches and proteins form local "circuits", which are part of still larger "networks" that govern the development of complex structures. Animal architecture is a product of genetic regulatory network architecture."

He goes on to say "another class of toolkit members belong to so-called signaling pathways. Cells communicate with one another by sending signals in the form of proteins that are exported and travel away from their source. Those proteins then bind to receptors on other cells, where they trigger a cascade of events, including changes in cell shape, migration, the beginning or cessation of cell multiplication, and the activation or repression of genes."

To summarize, the key abstractions the cell architecture supports are:

4. The spelling out of computational workflow components as a stable sequence of patterns that accomplishes a specific purpose,
5. A parallel management workflow specification with another sequence of patterns that assures the successful execution of the system's purpose (the computing network) and
6. A signaling mechanism that controls the execution of the workflow for gene expression (the regulatory network)
7. Real-time monitoring and control to execute genetic transactions which provide the self-* properties

Human Networks and Distributed Computing

Another example of a distributed system is the human network which has developed sophisticated abstractions and complex patterns for individuals to collaborate together and accomplish a common objective. The human networks are considered intelligent because they accomplish their goals in multiple ways using information collected from the external world and using it to control it. Just as cellular organisms have developed evolutionary best practices and pass them on from survivors to their successors, humans have evolved organizational best practices to leverage their individual capabilities as part of a group by defining and accomplishing common goals with high resiliency, effectiveness and scaling. A human network consists of a group of individuals organized to communicate and collaborate globally to carry out individual tasks executing a part of a distributed transaction using local intelligence and resources. Each group has a purpose and can be part of a larger group with its purpose consistent with that of the larger group. Each individual contributes (taking advantage of specialization) to the overall execution of the distributed transaction (implementing separation of concerns) as a part of the workflow as a managed directed acyclic graph. The group implements both the node-level and the network-level FCAPS management giving it the self-* management capabilities.

The effectiveness of the human network depends on the connections, communication and mastery (or specialization) of the individual human object. Better the quality of mastery of the individual node, the quality of connection and communication, higher the effectiveness. Humans have created organizational frameworks through evolution. According to Malone [13], organization consists of connected "agents" accomplishing results that are better than if they were not connected. An organization establishes goals, segments the goals into separate activities to be performed by different agents, and connect different agents and activities to accomplish the overall goals. Scalability is accomplished through hierarchical segmentation of activities and specialization.

There is always a balance between the cost of coordination of the agents and economies of scale obtained from increasing the network size which defines the nature of the connected network. Efficiency of the organization is achieved through specialization and segmentation. On the other hand agility of an organization depends on how fast the organization can respond to changes required to accomplish the goals by reconfiguring the network. Dynamic reconfiguration is accomplished using signaling abstractions such as addressing, alerting, supervision and mediation.

Both efficiency and agility are achieved through a management framework that addresses FCAPS of all network elements (in this case the agents). Project management is a specific example where Fault, configuration, accounting, performance and security are individually managed to provide an optimal network configuration with a coordinated work-flow. Functional organizations, and hierarchical and matrix organizational structures are all designed to improve the resiliency, efficiency, scaling and agility of an organization to accomplish the goals using both FCAPS management and signaling.

Connection management is achieved through effective communications framework. Over time, human networks have evolved various communications schemes and signaling forms the fundamental framework to configure and reconfigure networks to provide the agility. There are four basic abstractions that comprise signaling:

1 Alerting,
2 Addressing,
3 Supervision and
4 Mediation

Organizational frameworks are designed to implement these abstractions using distributed object management in human networks. Signaling allows prioritization of the network objectives and allocates resources in the form of distributed agents to accomplish the objectives and provides management control to resolve contention and mitigate risk. Elaborate workflows are implemented using the signaling mechanism to specialize and distribute tasks to various agents. The agents are used to collect information, analyze it and execute controls while organizing themselves as a group to accomplish the required goals.

Thus organizational hierarchies, project management, process implementation through workflows are all accomplished through the network object model with FCAPS abstractions and signaling. It is important to note that the signaling abstractions, while commonly used, have not been discussed widely in the distributed computing domain. First clear articulation is found in the description of SS7 signaling in telecommunications domain and a reference to it by Gartner group [14].

Telecommunications and Distributed Computing

For much of its history, AT&T and its Bell System functioned as a legally sanctioned, regulated monopoly. The fundamental principle, formulated by AT&T president Theodore Vail in 1907, was that the telephone by the nature of its technology would operate most efficiently as a monopoly providing universal service. Vail wrote in that year's AT&T Annual Report [15] that government regulation, "provided it is independent, intelligent, considerate, thorough and just" was an appropriate and acceptable substitute for the competitive marketplace." From the beginning of **AT&T** to today's remaking of **at&t**, much has changed but two things that remain constant are the universal service (on a global scale) and the telecom grade "trust" (providing reliable, secure and high performance connection at a reasonable cost) that are taken for granted. The Plain Old Telephone System (POTS) altered the communication landscape by connecting billions of humans anywhere any time at a reasonable cost. It provided the necessary managed infrastructure to create the voice dial tone, deliver it on demand and assure the connection to meet varying workloads and individual preferences with high availability, optimal performance and end-to-end connection security. The service assurance set a standard known as "telecom grade trust". Two major factors that contributed to the telecom grade trust are the end-to-end network management of various elements and the signaling network [14] that is used to dynamically manage the resources for every connection based on profiles.

Entry for FCAPS in Wikipedia, [16] states that it "is the ISO Telecommunications Management Network model and framework for network management. FCAPS is an acronym for fault, configuration, accounting, performance, and security which are the management categories into which the ISO model defines network management tasks. In non-billing organizations, accounting is sometimes replaced with administration. The comprehensive management of an organization's information technology (IT) infrastructure is a fundamental requirement. Employees and customers rely on IT services where availability and performance are mandated, and problems can be quickly identified and resolved. Mean time to repair (MTTR) must be as short as possible to avoid system downtimes where a loss of revenue or lives is possible".

All intelligent telecommunication network elements today are FCAPS enabled and Operation Support Systems are designed to provide day to day operations and management. The operation support systems and the network elements utilize the signaling abstractions to provide an elaborate communication infrastructure that enables collection, analysis and control of various elements to accomplish the business goals.

Similarly, the network infrastructure that forms the backbone of the Internet (consisting of the servers, routers, switches and other network elements) has liberally borrowed the FCAPS and signaling abstractions to implement agility and resiliency required. In fact FCAPS awareness has become a mandatory requirement to be a network element in both telecommunications and Information Technology infrastructure. As an example, as storage also became networked, various storage network elements have started to become FCAPS aware and participate in a common management framework.

Client Server, Peer-to-Peer, Grid and Cloud Computing Management

Resiliency, efficiency and scaling of a distributed system are very much dependent on the division of responsibilities between individual computing nodes, the placement of them, and connectivity between the nodes. In the SPC architecture, as we discussed earlier, the computational workflow and the management workflow are distributed among the various nodes. An SPC node provides an atomic computing unit and is programmed to perform useful activity with well-defined responsibility and interact with each other using a communication channel. Different placements of the responsibilities and communication schemes define various derived computing models [3]. We will discuss some of these models with respect to their resiliency, efficiency and scaling characteristics.

Client–Server and Peer-to-Peer Models

Each node in this model is a physical container managed by a single operating system to allocate resources to various hosted processes executing different tasks contributing to either the computing workflow or management workflow constituting a distributed transaction. The container may be a server providing services or a client receiving the services or both. Over a period of time, TCP/IP has become the standard for communication between processes located on different containers. Within the container the processes communicate via high speed shared memory or the PCI bus depending on the physical architecture. In the client server model, a server can be a client or a client a server depending on the context.

Services are implemented using multiple processes supported by the local operating system which communicate and collaborate with each other to implement the distributed transactions. Various FCAPS management workflows are implemented both using code that is mixed with the service workflow and separate processes in various nodes. For example replication is used to improve fault tolerance, performance and availability.

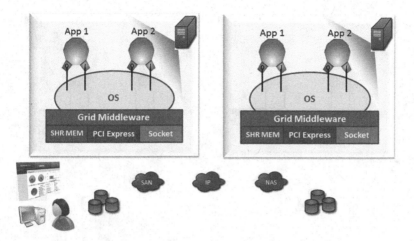

Fig. 2.6 Grid Computing Model

In peer-to-peer model, applications as processes in the operating system play a similar role, interacting cooperatively as peers to perform distributed transactions or computations without any distinction between clients and servers. The code in the peer processes also provides FCAPS management by maintaining consistency of application-level resources and synchronizes application-level actions when required.

Grid and Cloud Computing

Grid computing is designed to share disparate, loosely coupled IT resources across organizations and geographies. Using a grid middleware, the IT resources are offered as services. These resources include almost any IT component—computer cycles, storage spaces, databases, applications, files, sensors, or scientific instruments. Resources can be shared within a workgroup or department, across different organizations and geographies, or outside the enterprise.

Figure 2.6 depicts the role of Grid middleware and associated management software. The resources can be dynamically provisioned to users or applications that need them on demand.

With the advent of virtualization technologies, the physical enclosure that supported one operating system now allows multiple operating systems.

Figure 2.7 shows the new cloud computing model where multiple managed virtual servers contained in a physical enclosure can be dynamically provisioned, an operating system installed in every virtual server and applications can be run to share the CPU, network and storage resources.

Grid and cloud computing advances attempt to compensate for the deficiencies of the resource-centric management systems of conventional computing at the

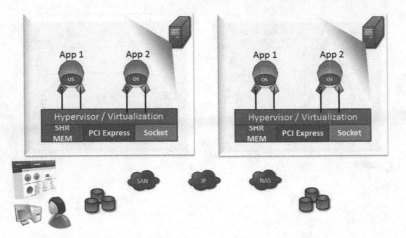

Fig. 2.7 Cloud computing model

application level by creating additional layers of resource utilization monitoring and management [17, 18]. Resulting automation and end-to-end visibility offered through collaborating management systems compensate for the inadequacy of the individual node operating system to see the global resource utilization involved in the distributed transaction. The distributed transaction resiliency, efficiency and scaling are improved through the control of end-to-end resources involved in the transaction.

Hardware Upheaval and The von Neumann Bottleneck

According to András Vajda [19], "The chip industry has recently coined the 'new' Moore's law, predicting that the number of cores per chip will double every two years over the next decade, leading us to the world of many-core processors-loosely defined as chips with several tens but more likely hundreds, or even thousands of processor cores." The reason for this shift from increasing clock speed to improve performance to increasing the number of cores in the same chip is two-fold:

1. Around the year 2004, the single processor clock-speed reached a maximum at about 4 GHz even though, the Moore's law allowed more transistors in smaller areas. The chip vendors decided to address this problem by increasing the number of cores in a chip thus improving the overall computing power.
2. This had the added advantage of saving power consumption and improving power dissipation, required to support higher speeds.

As the capabilities and speed of cores kept improving, the memory access speed lagged behind which led to the implementation of hardware multi-threading, a mechanism through which a core could support multiple thread contexts in hardware (including program counter and register sets, but sharing for example the cache memory) and fast switching between hardware threads stalled due to high latency operations. In addition, the many-cores in a chip also provide high-bandwidth interconnect which links multiple cores together to provide a single logical processing unit together.

The hardware advances implementing a large network of cores in a chip and a large network of chips in a server (often connected by a high-speed PCIExpress bus) create a new challenge to the current operating systems, and software practices that have their origins from the von Neumann serial computing model. The many-core architecture pushes the network of computing nodes inside a single enclosure with far higher bandwidths and faster access to stored program instructions than is possible by connecting different devices as is the state-of-the-art practice today in grid and cloud computing. This adds a new dimension in distributed transaction processing where resources now collaborate and cooperate inside the enclosure at high speed and the hardware multi-threading compounds the software challenge to take advantage of the inherent parallelism offered. There are three major reasons why the hardware upheaval unleashed by the many-core processors is making the current distributed software architectures to be reexamined:

1. The role of an operating system is to provide an abstraction layer of underlying hardware, taking care of interrupts, processor management, low-level interaction with peripherals, and allocate resources to applications with a unified interface for their use and management. Current generation operating systems which are node centric are not scalable because they do not have visibility of resources across the entire network. Figure 2.8 shows the recursive nature of networked SPC nodes with different resource allocation and management requirements. It is easy to see how the current device-centric OSs with their evolution from single-thread operating system data structures to support multi-threaded operating system data structures over time, with a large and complex code-base dealing with choosing correct lock granularity for performance, reasoning about correctness, and deadlock prevention are inadequate for network-centric resource management [20].

2. Application response time, in a many-core system depends on run-time workload fluctuations and latency constraints in a *shared processor and core network*-infrastructure as shown in Fig. 2.8. It therefore, becomes imperative to bring distributed transaction management infrastructure inside the many-core device allocating appropriate resources to various services that consume them based on business priorities, workload fluctuations and latency constraints. For example, if two computing nodes involved in collaboration with each other are in two different devices, the communication channel must be switched to socket communication, where as if they are communicating across two cores, shared memory would be an appropriate resource to be allocated. In a web based distributed transaction that spans across multiple geographies, the dynamic nature of the transaction demands dynamic resource allocation to optimally

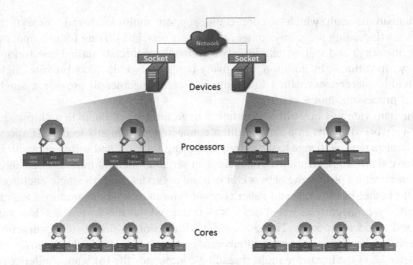

Fig. 2.8 The many-core servers and the recursive network-centric resource allocation and management requirements

execute the transaction. Current operating systems and management systems fall short in providing dynamic resiliency, efficiency and scaling for two reasons:

- The serial nature of von-Neumann computing node forces both service execution and service management to be intermixed as discussed above and introduces a level of complexity in dynamically managing resources globally based on changing business priorities, workload fluctuations and latency constraints.
- Current generation management systems have their origins in TCP/IP based narrow bandwidth environment and are not suited to leverage the high bandwidth and fast access to memory made available in many-core servers.

3. It is hard to imagine replicating current TCP/IP based socket communication, "isolate and fix" diagnostic procedures, and the multiple operating systems (that do not have end-to-end visibility or control of business transactions that span across multiple cores, multiple chips, multiple servers and multiple geographies) inside the next generation many-core servers without addressing their shortcomings. In order to cope with the scaling issues and utilize many-core technologies effectively, next generation service architecture has to emulate the architectural resiliency of cellular organisms that tolerate faults and implement command and control structures which enable execution of self-configuring, self-monitoring, self-protecting, self-healing and self-optimizing (in short self-*) business processes. Figure 2.9 shows the need for improving the resiliency, efficiency and scaling of many-core software services and their management architecture.

It is clear that current approaches to resource management, albeit with automation, are not sensitive to the distributed nature of transactions and contention

Fig. 2.9 The resiliency, efficiency and scaling diagram with many-core hardware upheaval and the von Neumann Bottleneck

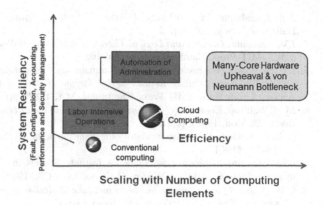

resolution of shared distributed resources, at best, is complex involving many layers of management systems. As von Neumann [4] pointed out, current design philosophy that "errors will become as conspicuous as possible, and intervention and correction follow immediately" does not allow scaling of services management with increasing number of computing elements involved in the transaction. Comparing the computing machines and living organisms, he points out that the computing machines are not as fault tolerant as the living organisms. More recent efforts, in a similar vein, are looking at resiliency borrowing from biological principles [21] to design future Internet architecture.

In the next chapter, we will revisit the design of distributed systems with a new non-von Neumann computing model (called Distributed Intelligent Managed Element (DIME[1]) Network computing model) that integrates computational workflows with a parallel implementation of management workflows to provide dynamic real-time FCAPS management of distributed services and end-to-end service transaction management.

The DIME network architecture provides a new direction to harness the power of many core servers with the architectural resiliency of cellular organisms and a high degree of scaling and efficiency. The DIME network architecture was first presented in WETICE 2010 in Larissa, Greece based on the workshop discussions started in WETICE 2009 in Groningen, Netherlands [22–24].

References

1. K.P. Birman, *Reliable Distributed Systems: Technologies, Web Service and Applications* (Springer, NY, 2005), p. 19
2. G. Coulouris, J. Dollimore, T. Kindberg, *Distributed Systems, Concepts and Design*, 3rd edn. (Addison Wesley, New York, 2001), p. 1

---

[1] DIME[TM], Cloud-DNA and Dime Network Architecture are Trade Marks of Kawa Objects Inc.

3. A.S. Tanenbaum, M. van Sheen, *Distributed Systems, Principles and Paradigms* (Prentice Hall, New Jersey, 2002), p. 2

4. J.V. Neumann, *General and Logical Theory of Automata*, William Aspray and Arthur Burks (eds.), (MIT Press, Cambridge, 1987), p. 408

5. J.V. Neumann, Papers of John von Neumann on Computing and Computer Theory, in *Charles Babbage Institute Reprint Series for the History of Computing*, ed. by William Aspray, Arthur Burks (MIT Press, Cambridge, MA, 1987), pp. 409–474

6. M.M. Waldrop, *Complexity: The Emerging Science at the Edge of Order and Chaos* (Penguin Books, London, 1992), p. 218

7. F. Jacob, J. Monod, Genetic regulatory mechanisms in the synthesis of proteins. J. Mol. Biol. **3**, 318–356 (1961)

8. C.G. Langton, *Artificial Life*, Santa Fe Institute Studies in the Sciences of Complexity, in Proceedings **6**. (Addison Wesley, Redwood City, CA, 1989)

9. G.B. Dyson, *Darwin among the Machines, the evolution of global intelligence* (Addition Wesley, Reading, MA, 1997), p. 123. Helix Books

10. R. Dawkins, *The Selfish Gene* (Oxford University Press, New York, 1989), p. 23

11. Y. Sako, (2010), http://www.researchsea.com/html/article.php/aid/5501/cid/1/research/studying_cell_signaling_using_singlemolecule_imaging.html?PHPSESSID=rlufujxhvtg. (Accessed 27 November 2010) from Asia Research News: http://www.researchsea.com/html/article.php/aid/5501/cid/1/research/studying_cell_signaling_using_singlemolecule_imaging.html?PHPSESSID=rlufujxhvtg

12. S.B. Carroll, *The New Science of Evo Devo—Endless Forms Most Beautiful* (W. W. Norton & Co, New York, 2005), pp. 12 ,106 ,113, 129

13. T.W. Malone, Organizing Information Processing Systems: Parallels Between Human Organizations and Computer Systems, in *Cognition, computing, and cooperation*, ed. by S.P. Robertson, W. Zachary, J.B. Black (Ablex Publishing, New Jersey, 1990)

14. Gartner Dataquest Says Next Generation Signaling to Prosper with Emergence of Next Generation Networks. Business Wire. Accessed 12 February 2001. (http://www.allbusiness.com/technology/software-services-applications-information/6029582-1.html)

15. AT&T, Brief history (2010), http://www.corp.att.com/history/history3.html. Accessed 28 November 2010 from AT&T history

16. Wikipedia: http://en.wikipedia.org/wiki/FCAPS (Retrieved 11 29, 2010)

17. R. Buyya, R. Ranjan, Special section: Federated resource management in grid and cloud computing systems. Fut. Gen. Comput. Syst. **26**, 1189–1191 (2010)

18. R. Buyyaa, C.S. Yeoa, S. Venugopala, J. Broberga, I. Brandicc, Cloud computing and emerging IT platforms: Vision, hype, and reality for delivering computing as the 5th utility. Fut. Gen. Comput. Syst. **25**(6), 599–616 (2009)

19. A. Vajda, *Programming Many-Core Chips* (Springer, New York, 2011), p. 3

20. D. Wentzlaff, A. Agarwal, Factored operating systems (fos): the case for a scalable operating system for multicores. SIGOPS Oper. Syst. Rev. **43**(2), 76–85 (2009)

21. S. Balasubramaniam, K. Leibnitz, P. Lio', D. Botvich, M. Murata, Biological Principles for Future Internet Architecture Design. IEEE Comm. Mag. **49**(7), 44 (2011)

22. P. Goyal, The Virtual Business Services Fabric: An Integrated Abstraction of Services and computing Infrastructure, in *Proceedings of WETICE 2009: 18th IEEE International Workshops on Enabling Technologies: Infrastructures for Collaborative Enterprises*, 2009, pp. 33–38

23. P. Goyal, R. Mikkilineni, M. Ganti, Manageability and Operability in the Business Services Fabric, in *18th IEEE International Workshops on Enabling Technologies: Infrastructures for Collaborative Enterprises, 2009. WETICE'09*, 29 June–1 July 2009, pp. 39–44

24. P. Goyal, R. Mikkilineni, M. Ganti, FCAPS in the Business Services Fabric Model, in *18th IEEE International Workshops on Enabling Technologies: Infrastructures for Collaborative Enterprises, 2009. WETICE'09*, 29 June–1 July 2009, pp. 45–51

Chapter 3
Distributed Intelligent Managed Element (DIME) Network Architecture Implementing a Non-von Neumann Computing Model

Abstract A new computing model called Distributed Intelligent Managed Element incorporates fault, configuration, accounting, performance and security (FCAPS) management using a signaling network overlay and allows the dynamic control of a set of distributed computing elements in a network. Each node is a computing entity (a Turing machine implemented using von Neumann computing model) modified by endowing it with self-management and signaling capabilities to collaborate with similar nodes in a network. The separation of parallel computing and management channels allows the end to end transaction management of computing tasks (provided by the autonomous distributed computing elements) to be implemented as network-level FCAPS management.

The DIME Network Architecture and the Anatomy of a DIME

"Our notion that Turing machines represent the basis for our current view of cognition is completely off-track"—Louise Barrett, "Beyond the Brain: How Body and Environment Shape Animal and Human Minds", Princeton University Press, Princeton, NJ, 2011, p. 121.

The raison d'etre for Distributed Intelligent Managed Element (DIME) computing model is to fully exploit the parallelism, distribution and massive scaling possible with multi-core and many-core processor based servers, laptops and mobile devices supporting hardware assisted virtualization and create a computing architecture in which the services and their management in real-time are decoupled from the hardware infrastructure and its management. However, the model lends itself to be implemented (i) from scratch to exploit the many core servers and (ii) in current generation servers exploiting multi-thread computing features available in current operating systems such as Linux and Windows.

R. Mikkilineni, *Designing a New Class of Distributed Systems*, SpringerBriefs in Electrical and Computer Engineering, DOI: 10.1007/978-1-4614-1924-2_3,
© The Author(s) 2011

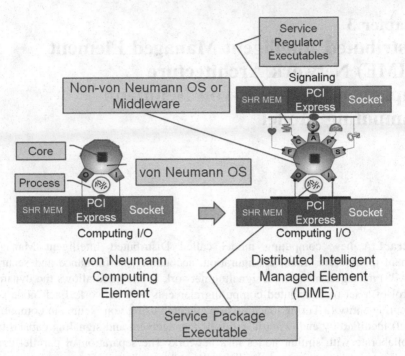

Fig. 3.1 The von-Neumann and DIME computing models. For a description of the DIME network architecture and the genetic transactions, please see the video http://youtu.be/Ft_W4yBvrVg

Borrowing the fault, configuration, accounting, performance and security (FCAPS) management and signaling abstractions from cellular organisms and human networks, the DIME computing model [1] exploits the parallelism to implement a signaling network overlay over a network of von Neumann SPC computing nodes. The computing node is either a core in a many-core server, a process in a conventional operating system, or a processor in any mobile device or a laptop. Multiple threads available in each core or an operating system process implementation are exploited to implement a self-managed computing element called the DIME. Each DIME presents a computing element that can execute a managed computing process with fault, configuration, accounting, performance and security management. Figure 3.1 shows a comparison between the von Neumann SPC computing model and the DIME computing model. The DIME network computing model exploits the multithread capability offered in the computing element (either as a process in a conventional operating system or as a core in a many-core system) to separate management and computing threads. The parallelism is exploited to implement the management of a Turing machine. The parallel signaling network allows the management of a network of managed Turing nodes. The recursive network composition model is ideally suited to implement recursive state machines and thus implement service workflows.

The parallelism of service execution and service regulation allows real-time monitoring of service behavior and control, based on policies and constraints specified by the regulators both at the node level and at the network level. The DIME network architecture thus allows the description and management of the service to be separated from the execution of the service (using a computing thread called Managed Intelligent Computing Element (MICE)). The signaling control network allows parallel management of the service workflow. In Step 1, the service regulator instantiates the DIME and provisions the MICE based on service specification. In Step 2, the MICE is loaded, executed, and managed by the service regulation policies. At any time, the MICE can be controlled through its FCAPS management mechanism by the service regulator.

There are three key features in this model that differentiate it from all other models:

1. The self-management features of each SPC node with FCAPS management using parallel threads allow autonomy in controlling local resources and provide services based on local policies. Each node keeps its state information and history of its transactions. The DIME node provides managed computing services, using the MICE to other DIMEs based on local and global policies.
2. The network aware signaling abstractions allow a group of DIMEs to be programmed to manage themselves with sub-network/network level FCAPS management based on group policies and execute a service workflow as a managed directed acyclic graph (DAG).
3. Run-time profile based FCAPS management (at the group level and at the node level) allows a composition scheme by redirecting the MICE I/O to provide recombination and reconfiguration of service workflows dynamically at run-time.

The MICE provides the logical type that performs everything that is feasible within that logical type (a Turing machine) and the DIME FCAPS management provides a higher logical type (management of the Turing machine) which describes and controls what is feasible in the MICE [2]. These features provide the powerful genetic transactions namely, replication, repair, recombination and reconfiguration that have proven to be essential for the resiliency of cellular organisms [3].

The self-management of the DIME and the task execution (using the MICE) are performed in parallel using the stored program control computing devices. Figure 3.2 shows the anatomy of a DIME. Each DIME is implemented as group of multi-process, multi-thread components, as shown in Fig. 3.2.

The DIME orchestration template provides the description for instantiating the DIME using an SPC computing device with appropriate resources required (CPU, memory, network bandwidth, storage capacity, throughput and IOPs). The description contains the resources required, the constraints and the addresses of executable modules for various components and various run time commands the DIME obeys. This description is called the regulatory gene and contains all the information required to instantiate the DIME with its FCAPS management

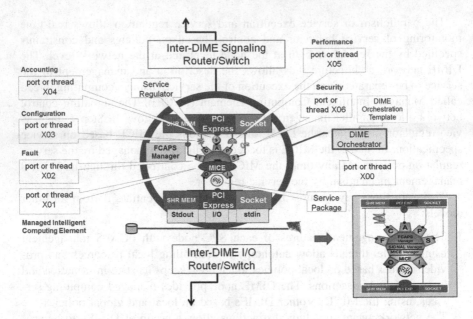

Fig. 3.2 The anatomy of a DIME using parallelism and multi-threading capabilities

components, the MICE and the signaling framework to communicate with external DIME infrastructure.

The configuration commands provide the ability for the MICE to be set up with appropriate resources and I/O communication network to be set up to communicate with other DIME components to become a node in a service delivery network implementing a workflow.

Figure 3.3 shows the service implementation with a service regulator and the service execution package [4].

Signaling allows groups of DIMEs to collaborate with each other and implement global policies with high degree of agility that the parallelism offers. The signaling abstractions are:

1. Addressing: For network based collaboration, each FCAPS aware DIME must have a globally unique address and any services platform using DIMEs must provide name service management.
2. Alerting: Each DIME is capable of self-identification, heartbeat broadcast, and provides a published alerting interface that describes various alerting attributes and its own FCAPS management.
3. Supervision: Each DIME is a member of a network with a purpose and a role. The FCAPS interfaces are used to define and publish the purpose, role and various specialization services that the DIME provides as a network community member. Supervision allows contention resolution based on roles and purpose. Supervision also allows policy monitoring and control.

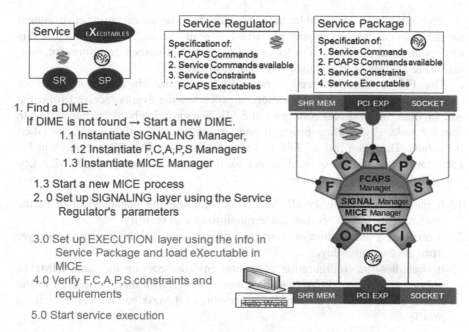

1. Find a DIME.
   If DIME is not found → Start a new DIME.
       1.1 Instantiate SIGNALING Manager,
       1.2 Instantiate F,C,A,P,S Managers
       1.3 Instantiate MICE Manager

   1.3 Start a new MICE process
2. 0 Set up SIGNALING layer using the Service
   Regulator's parameters

3.0 Set up EXECUTION layer using the info in
    Service Package and load eXecutable in
    MICE
4.0 Verify F,C,A,P,S constraints and
    requirements

5.0 Start service execution

Fig. 3.3 The service regulator and service package implementation using a DIME

4. Mediation: When the DIMEs are contending for resources to accomplish their specific mission, or require prioritization of their activities, the supervision hierarchy is assisted with mediation object network that provides global policy enforcement.

The DIME local Manager (DLM) sets up the other DIME components; it monitors their status and manages their execution based on local policies. Upon a request to instantiate a DIME, the DLM, based on the role assumed by the DIME, sets up and starts three independent threads to provide the Signaling Manager (SM), the MICE Manager (MM) and the FCAPS Manager (FM) functions.

The SM is in charge of the "signaling channel". It sends or receives commands related to the management and setting up of DIME to guarantee a scalable, secure, robust and reliable workflow execution. It also provides inter-DIME switching and routing[1] functions.

The MM is a passive component which starts an independent process, the MICE, which in turn executes the task (or tasks), assigned to that DIME and, on completion, notifies the event to the SM. All the actions related to the task execution, which are performed by the MICE including memory, network, and storage I/O, are parameterized and can be configured and managed dynamically by the SM through

---

[1] Each DIME is globally addressable and supports network connectivity for both signaling and computing workflows using inter-DIME routing and switching.

the FM via the MM. This enables both the ability to set up the execution environment on the basis of the user requirements and, overall, the ability to reconfigure this environment, at run-time, in order to adapt it to new, foreseen or unforeseen, conditions (e.g. faults, performance and security conditions).

The FM is the "connection point" between the channels for workflow management and workflow execution. It processes the events received from the SM or from the MM and configures the MICE appropriately to load and execute specific tasks (by loading program and data modules available from specified locations). The main task of FM is the provisioning of FCAPS management for each task loaded and executed in the local DIME. This makes the FM a key component of the entire system:

1. It handles autonomously all the issues regarding the management of faults, resources utilization, performance monitoring and security,
2. It provides a "separation of concerns" which decouples the management layer from the execution layer,
3. It simplifies the configuration of several environments on the same DIME to provide appropriate FCAPS management of each task that is assigned to the MICE which in turn, performs the processing of the task based on an associated profile.

Not all the components seen above have to be active, at the same time: the DLM will start only the components that are required to accomplish the functionalities specified by the role of each DIME in the network.

DIME Network Architecture and the Architectural Resiliency of Cellular Organisms

The DIME network architecture supports the genetic transactions of replication, repair, recombination and rearrangement. Figure 3.4 shows a single node execution of a service in a DIME network.

A single node of a DIME can execute a workflow by itself. Instantiating a sub-network provides a way to implement a managed DAG executing a workflow. Replication is implemented by executing the same service as shown in Fig. 3.5.

By defining service S1 to execute itself, we replicate S1 DIME. Note that S1 is a service that can be programmed to terminate instantiating itself further when resources are not available. In addition, dynamic FCAPS (parallel service monitoring and control) management allows changing the behavior of any instance from outside (using the signaling infrastructure) to alter the service that is executed.

The ability to execute the control commands in parallel allows dynamic reconfiguration or replacement of services during run time. For example by stopping service S1 and loading and executing service S2, we dynamically change

Fig. 3.4 Single node
execution of a DIME

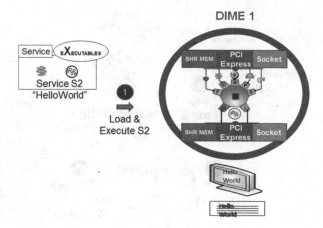

Fig. 3.5 DIME replication

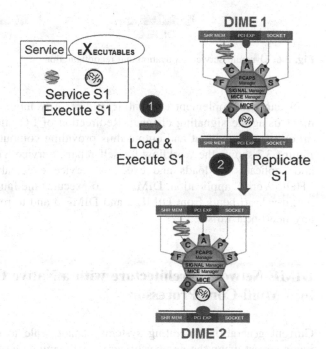

the service during run time. We can also redirect I/O dynamically during run time.
Any DIME can also allow a sub-network instantiation and control as shown

in Fig. 3.6. The workflow orchestrator instantiates the worker nodes, monitors
heartbeat and performance of workers and implement fault tolerance, recovery,
and performance management policies.

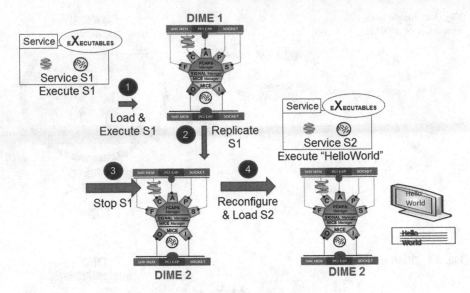

Fig. 3.6 Dynamic service replication and reconfiguration

It can also implement accounting and security monitoring and fault management using the signaling channel. Redirection of I/O allows dynamic reconfiguration of worker input and output thus providing computational network control. Figure 3.7 shows the workflow for self-repair. Service S1, instantiates service S2 and replicates it, loads and executes service executables from S2 executing "HelloWorld" application. DIME 1 also executes the fault management policy to monitor heart-beats from DIME 2 and DIME 3 and to re-instantiate a DIME 4 if any heart-beat fails.

DIME Network Architecture with a Native OS in a Multi-Core Processor

Current generation Operating systems cannot scale to encompass the resource management when the number of cores in a many core server reaches a threshold dictated by mechanisms choosing correct lock granularity for performance, reasoning about correctness, and deadlock prevention. The impact of the *operating system gap (the difference between the number of cores available in a server and the number of cores visible to a single instance of the operating system deployed in it)* is dramatic when you consider current deployment scenarios. In one instance, a 500 core server is used as 250 dual core servers with 250 Linux images. In this case, in spite of proximity and high bandwidth, the TCP/IP based socket

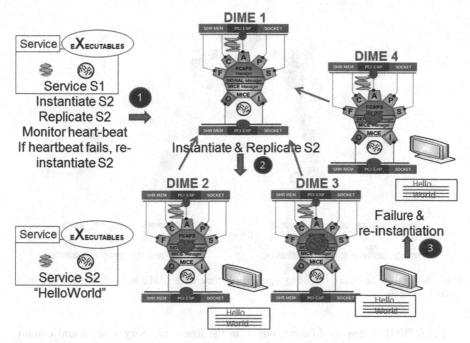

Fig. 3.7 Self-repair workflow

abstractions limit the performance by not utilizing the hardware resources and parallelism made available with new many-core architectures.

Using the DIME architecture we propose a new scheme in which a native operating system is implemented that converts each core into a DIME and provides inter-DIME and intra-DIME signaling capability to implement network-wide FCAPS management of service workflows. Figure 3.8 shows the DIME implementation in multi-core servers [5, 6].

A native OS called Parallax is implemented, to demonstrate feasibility, using the assembler language at the lowest level for efficiency and provides a C/C ++ programming API for higher level programming. It is implemented to execute on 64-bit multi-core Intel processors. Each core is encapsulated as a DIME addressable as a network element with its own FCAPS management module. Each DIME has two communications channels supported by Ethernet, one for signaling and another for MICE I/O communications. The signaling Channel is used to execute FCAPS commands and change FCAPS parameters at run time. The data channel is dynamically reconfigurable to set up inter-MICE communications, I/O paths and network and storage paths. This allows a composition scheme for creating a network of MICEs to execute a DAG very similar to using PIPE in UNIX but while the applications are running. The kernel provides memory management, CPU resource management, storage/file management, network routing and switching functions, and signaling management.

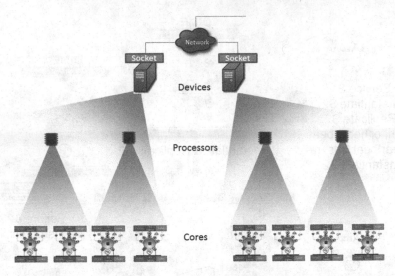

Fig. 3.8 Parallax (a native operating system) implementing DIME network architecture in a multi-core server

Each DIME maps to different pages in the linear memory system and cannot access pages to which it is not assigned. Security is provided at the hardware level for this memory protection. Once a program has completed its execution, all memory that it had in use is returned to the system. Limits can be set on how much memory each DIME is able to allocate for itself. Memory is divided into a shared memory partition where the Parallax Kernel resides and partitions that are devoted to each core. Memory can be dynamically adjusted on each core on as needed basis in 2 MiB chunks. Memory allocated to a DIME (core) can only be accessed by that DIME based on its security configuration. With dedicated resources, each DIME can be viewed as its own separate computing entity. If a DIME completes its task and is free, it is given back to the pool of available resources. The network management assures discovery and allocation of available DIMEs in the pool for new tasks. The signaling allows addressability at the thread level.

Parallax offers local storage per server (Shared with each DIME within the system) as well as centralized file storage shared via the Orchestrator between all servers. Booting the OS via the network is also a possibility for systems that do not need permanent storage or for cost saving measures.

Under Parallax, all network communication is done over raw Ethernet frames. Conventional operating systems use TCP/IP as the main communication protocol. By using raw packets we have created a much simpler communication framework as well as removed the overhead of higher-level protocols, thereby increasing the maximum throughput. The use of raw Ethernet packets has already seen great success with the ATAoE protocol invented by Co-Raid for use in their network storage devices. Eventually, PCIExpress, and TCP/IP will be added along with Shared Memory.

Under Parallax, each DIME is addressable as a separate entity via the signaling and data channels. With the signaling layer, program parameters can be adjusted during run-time. DIMEs have the ability to communicate with other DIMEs for co-operation. The Orchestrator, from which the policies are implemented, communicates with the DIMEs for the purpose of coordination and control. Instruction types can be directly encoded into the 16-bit Ether-Type field of each Ethernet frame shown in Fig. 3.2. By making use of the EtherType field for specific purposes we can streamline the way in which packets are routed within a system. Packets can be tagged as Signaling/Data packets as well as whether or not they are destined for the overall system or rather just a specific DIME.

The proof-of-concept prototype system consists of three components:

1. A service component development program that takes assembler or C/C++ programs and compiles them to be executed on an Intel Xeon multi-core Servers.
2. Parallax Operating System that is used to boot the servers with Intel Xeon cores and create the DIME Network with each core acting as a DIME. The DIME allows dynamic provisioning of memory for each DIME. It supports executing multiple threads concurrently to provide DIME FCAPS management over a signaling channel. It enables fault management by broadcasting a heartbeat over the signaling network. It allows loading, executing, and stopping an executable on demand. It supports DIME discovery through signaling channel.
3. A run-time service orchestrator that allows DIME network management.

Figure 3.9 shows the Proof-of-concept set up using three servers with Intel-Xeon processors where a DIME networks is deployed and various features such as discovery, service scaling, fault management and dynamic reconfiguration are demonstrated.

There are parallel efforts that are underway to architect a new OS for many-core servers:

1. Tessellation [7]: It is predicated on two central ideas: Space–Time Partitioning (STP) and Two-Level Scheduling. STP provides performance isolation and strong partitioning of resources among interacting software components, called Cells. Two-Level Scheduling separates global decisions about the allocation of resources to Cells from application-specific scheduling of resources within Cells.
2. Barrellfish [8]: It uses a multi-kernel model which calls for multiple independent OS instances communicating via explicit messages. Barrelfish factors the OS instance on each core into a privileged-mode CPU driver and a distinguished user-mode monitor process. CPU drivers are purely local to a core, and all inter-core coordination is performed by monitors. The distributed system of monitors and their associated CPU drivers encapsulate the functionality found in a typical monolithic microkernel such as scheduling, communication, and low-level resource allocation. The rest of Barrelfish consists of device drivers and system services (such as network stacks, memory allocators, etc.) which, run in user-level processes as in a microkernel. Device interrupts are

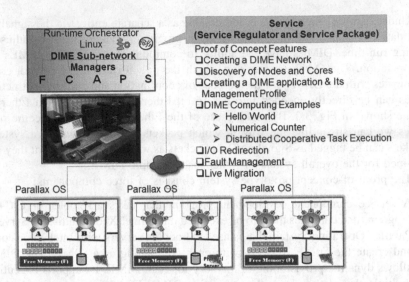

Fig. 3.9 The proof-of-concept setup. A video of the demo is available at http://www.youtube.com/ watch?v=y-0R-cRLFsk

routed in hardware to the appropriate core, de-multiplexed by that core's CPU driver, and delivered to the driver process as a message.

3. Factored Operating System (FOS) [9]: According to the authors, "FOS is a new operating system targeting many-core systems with scalability as the primary design constraint, where space sharing replaces time sharing to increase scalability. We describe FOS which is built in a message passing manner, out of a collection of Internet inspired services. Each operating system service is factored into a set of communicating servers which in aggregate implement a system service. These servers are designed much in the way that distributed Internet services are designed, but instead of providing high level Internet services, these servers provide traditional kernel services and replace traditional kernel data structures in a factored, spatially distributed manner. FOS replaces time sharing with space sharing. In other words, FOS's servers are bound to distinct processing cores and by doing so do not fight with end user applications for implicit resources such as TLBs and caches." They suggest redesigning traditional OSs using their approach for scalability.

4. Helios [10]: Helios is an operating system designed to simplify the task of writing, deploying, and tuning applications for heterogeneous platforms. Helios introduces satellite kernels, which export a single, uniform set of OS abstractions across CPUs of disparate architectures and performance characteristics. Access to I/O services such as file systems are made transparent via remote message passing, which extends a standard microkernel message-passing abstraction to a satellite kernel infrastructure. Helios retargets applications to available ISAs by compiling from an intermediate language. The authors

compare their approach to Barrelfish as follows: Barrelfish focuses on gaining a fine-grained understanding of application requirements when running applications, while the focus of Helios is to export a single-kernel image across heterogeneous coprocessors to make it easy for applications to take advantage of new hardware platforms."

5. Corey [11]: The authors argue that applications should control sharing: "the kernel should arrange each data structure so that only a single processor need update it, unless directed otherwise by the application. Guided by this design principle, this chapter proposes three operating system abstractions (address ranges, kernel cores, and shares) that allow applications to control inter-core sharing and to take advantage of the likely abundance of cores by dedicating cores to specific operating system functions. Measurements of micro-benchmarks on the Corey prototype operating system, which embodies the new abstractions, show how control over sharing can improve performance. Application benchmarks, using MapReduce and a Web server, show that the improvements can be significant for overall performance: MapReduce on Corey performs 25% faster than on Linux when using 16 cores. Hardware event counters confirm that these improvements are due to avoiding operations that are expensive on multicore machines."

All these efforts recognize the importance of application's requirements in controlling the resources and provide a way to mediate between the many-core resources and fluctuating application needs. All these approaches implement application services and the resource mediation services using the same serial von Neumann SPC model.

However, the DIME approach proposed in this chapter takes a different route to leverage the parallelism offered by multi-core and many-core architecture to implement the service management workflow as an overlay over the service workflow implemented over a network of SPC nodes. The separation and parallel implementation of the service regulation improve both the resilience, and the efficiency. The recursive (or fractal-like) network composition model eliminates the scaling limitation.

DIME Network Architecture Implementation in Linux

The DIME computing model offers a simple way to implement service virtualization independent of current generation virtualization technologies. One implementation [4] uses the multi-process, multi-thread support in the Linux operating system to implement the DIME network. By encapsulating a Linux based processes with parallel FCAPS management and providing a parallel signaling channel, this implementation demonstrates auto-scaling, self-repair, live-migration, performance management and dynamic reconfiguration of workflows without the need for a Hypervisor-based server virtualization.

Two points are worth noting about this implementation:

1. The workflow assigned to a DIME network consists of a set of tasks arranged in a DAG. Each node of this DAG contains both the task executables (which itself could be another DAG) and the profile DAG as a tuple < task (SP), profile (SR) >: in this way, it is possible not only to specify what a DIME has to do or execute but also its management (how this has to be done and under what constraints). These constraints allow the control of FCAPS management both at the node level and the sub-network level. In essence, at each level in the DAG, the tuple gives the blueprint for both management and execution of the down-stream graph. Under these considerations, it is easy to understand the power of the proposed solution in designing self-configuring, self-monitoring, self-protecting, self-healing and self-optimizing distributed service networks.

2. An ad hoc DIME network, with parallel signaling and computing workflows, is implemented using two classes of DIMEs. Signaling DIMEs responsible for the management layer at the network level are of the type Supervisor and the Mediator. The Supervisor sets up and controls the functioning of the sub network of DIMEs where the workflow is executed. It coordinates and orchestrates the DIMEs through the use of the Mediators. A Mediator is a specialized DIME for providing predefined roles such as fault or configuration or accounting or performance or security management. Worker DIMEs con-stituting the "execution" layer of the network perform domain specific tasks that are assigned to them. A worker DIME, in practice, provides a highly configurable execution environment built on the basis of the requirements/ constraints expressed by the developers and conveyed by the Service Regulator.

The deployment of DIMEs in the network, the number of signaling DIMEs involved in the management level, the number of available worker DIMEs and the division of the roles are established on the basis of the number and the type of tasks constituting the workflow and, overall, on the basis of the management profiles related to each task. The profiles play a fundamental role in the proposed solution; each profile, in fact, contains the indication about the control and the configuration of both the signaling layer and execution environment for setting up the DIME that will handle the related task. Figure 3.10 shows the DIMEs in Linux schematic.

Using signaling and FCAPS measurements, it is possible to identify each DIME's context and configure resources appropriately based on end to end transaction latency requirements (using shared memory, PCIExpress and socket communications), workload fluctuations and business priorities. The architectural innovation introduced here based on FCAPS and signaling abstractions radically transforms the Linux process implementation with a resiliency that surpasses current state of the art. For example, fault management, performance management, security management are implemented at both the process level using a self-managed DIME and at the DIME network level which assures service workflow FCAPS management that spans across multiple processes that are distributed. When two DIMEs reside in the same enclosure where shared memory is more effective, the communication is dynamically configured to support shared memory. When two DIMEs are separated

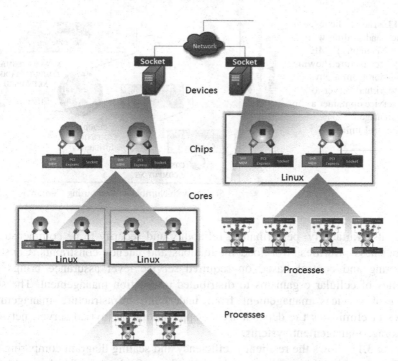

Fig. 3.10 Each process in Linux is encapsulated as a DIME with FCAPS management and signaling capability. In the many-core server, multiple images of Linux are deployed. In each image, the processes are encapsulated as DIMEs and inter-DIME communication is provided through shared memory, PCIExpress or socket based on requirement

by geography the communication is configured to support socket based TCP/IP. Figure 3.10 shows how multiple Linux images are injected with a middleware library called DIMEX which implements the DIME network architecture. The middleware allows executing current services/applications in a DIME or write new applications that exploit run-time FCAPS management to implement end-to-end distributed transaction management. In the next chapter, we will discuss various features demonstrated by the prototype.

Grids, Clouds, DIMEs and Their Management

The DIME Network architecture brings FCAPS management to a von Neumann computing element either by encapsulating a process in a conventional operating system or a core in a many-core server using a native operating system. The programmability and execution of management at the node level and at the network level using parallel signaling network provides a fine-grain end-to-end distributed transaction management. It is therefore possible to implement end-to-end resource management that contributes to a distributed transaction which

Fig. 3.11 The resiliency, efficiency and scaling with non-von Neumann DIME network architecture showing the transition from a physical server to virtual server to a virtual service container as an atomic managed computational unit

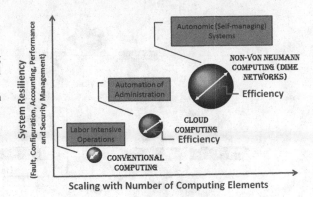

assures its availability, performance, reliability and security to be commensurate with business priorities, workload fluctuations and latency constraints. Constant monitoring and control based on required service level assurance brings the resiliency of cellular organisms to distributed transaction management. The separation of services management from underlying infrastructure management reduces or eliminates the dependence of applications on myriad server, network and storage management systems.

Figure 3.11 shows the resiliency, efficiency and scaling diagram comparing the DIME networks with conventional computing, Grid and Cloud computing architectures.

As the number of cores in a many-core server increase, while current OSs and various management systems (server management, virtual server management, network resource mediation systems and storage resource mediation systems) increase the complexity, the scaling of DIME network architecture allows many of the features offered by current virtualization technologies such as auto-scaling, self-repair, auto-performance management, live migration etc. without the need for complex hypervisor or other technologies.

Both signaling and service component network management allow a new way using service switching to provide FCAPS management which, heretofore has been provided by multiple resource management systems. The service-centricity as opposed to resource-centric management could offer simplicity of resource deployment with many-core server. When using a 500 or 1000 core server and using WAN connectivity between the servers, it would be unnecessary to use Storage Area Networks with Fibre channel inside the server or the data center. Similarly, with end-to-end transaction security management which controls reads and writes at every node, current Firewall and routing technologies need not be replicated inside the server. Figure 3.12 shows new many-core server architecture with DIME networks.

In the DIME network architecture, the multi-tenancy is transformed from the number of users that can be supported with FCAPS management in a container (A physical server in conventional computing and a virtual server in a cloud) to

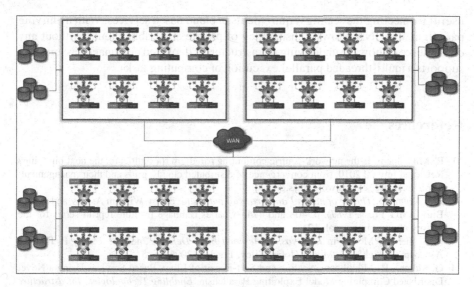

Fig. 3.12 A many-core Server WAN network. The signaling and FCAPS management at both the node level and the network level allows a simplification of service management by eliminating many of the current generation resource management automation systems and replacing it with services switching

number of service transactions that are supported with FCAPS management in a distributed set of enclosures. Each service transaction can be dynamically configured with assurance of FCAPS management of all the nodes that contribute to the transaction based on business priorities, workload fluctuations and latency constraints.

In a network-centric service switching architecture an end-to-end distributed transaction becomes a connection management task. For example all reads and writes are controlled by network level and node level policies. The service switching architecture brings features such as call waiting, call forwarding, call broadcast, and the 800 service call to manage the distributed transaction FCAPS service levels based on service profiles of both suppliers and consumers.

It is also important to note that while the hardware upheaval offers major cost savings in power and space savings alone, a transition from conventional computing with multi-tenancy at the physical server is improved by the multi-tenancy at the virtual server level by the number of virtual servers that can be run in a single enclosure. The DIME network architecture takes the scaling to the next higher level by the number of service transactions that can be supported in an enclosure when more resources have to be added.

In the next chapter we will discuss some applications of the DIME network architecture both in the short run where current generation hardware and software are transparently migrated and in the long run where a new class of distributed services are designed, deployed and assured using the new architecture. The latter is very useful for meeting the scaling requirements of cloud based services. With a unifying paradigm, the DNA allows transparency of private and public clouds without any dependence on how the underlying infrastructure is deployed or managed as long as it supports a multi-threaded parallel execution of computing tasks.

References

1. R. Mikkilineni, Is the network-centric computing paradigm for multicore, the next big thing? Retrieved July 22 2010, from convergence of distributed clouds, grids and their management: http://computingclouds.wordpress.com
2. J.V. Neumann, *Theory of natural and artificial automata*. ed. by William Aspray and Arthur Burks (MIT Press, 1987), p. 408 and 474. (Charles Babbage institute reprint series for the history of computing, vol. 12)
3. P. Stanier, G. Moore, in *Embryos, Genes and Birth Defects*, 2nd edn. ed. by P. Ferretti, A. Copp, C. Tickle, G. Moore (John Wiley, London, 2006), p. 5
4. G. Morana, R. Mikkilineni, Scaling and Self-Repair of Linux Based Services Using a Novel Distributed Computing Model Exploiting Parallelism. *Enabling Technologies: Infrastructure for Collaborative Enterprises (WETICE) 2011 20th IEEE International Workshops on*, pp. 98–103, 27–29 June 2011
5. R. Mikkilineni, I. Seyler, Parallax—A New Operating System for Scalable, Distributed, and Parallel Computing. *Parallel and Distributed Processing Workshops and Phd Forum (IPDPSW), 2011 IEEE International Symposium on*, pp. 976–983, 16–20 May 2011
6. R. Mikkilineni, I. Seyler, Parallax—A New Operating System Prototype Demonstrating Service Scaling and Self-Repair in Multi-core Servers. *Enabling Technologies: Infrastructure for Collaborative Enterprises (WETICE) 2011 20th IEEE International Workshops on*, pp. 104–109, 27–29 June 2011
7. J.A. Colmenares, S. Bird, H. Cook, P. Pearce, D. Zhu, J. Shalf, S. Hofmeyr, K. Asanovic, J. Kubiatowicz, Tesselation: space–time partitioning in a manycore client OS, in *Proceedings of the 2nd USENIX Workshop on Hot Topics in Parallelism (HotPar'10)* (Berkeley, June 2010)
8. A. Baumann, P. Barham, P.-E. Dagand, T. Harris, R. Isaacs, S. Peter, T. Roscoe, A. Schupbach, A. Singhania, The Multikernel: A New OS Architecture for Scalable Multicore Systems, in *Proceedings of the 22nd ACM Symposium on OS Principles*, Big Sky, October 2009
9. D. Wentzlaff, A. Agarwal, Factored operating systems (FOS): the case for a scalable operating system for multicores. SIGOPS Oper. Syst. Rev. **43**(2), 76–85 (2009)
10. E.B. Nightingale, O. Hodson, R. McIlroy, C. Hawblitzel, G. Hunt, Helios: heterogeneous multiprocessing with satellite kernels. ACM, SOSP'09, Big Sky, October 11–14, 2009
11. O. Mao, F. Kaashoek, R. Morris, A. Pesterev, L. Stein, M. Wu, Y. Dai, Y. Zhang, Z. Zhang, Corey: An Operating System for Many Cores, in *Proceedings of the 8th USENIX symposium on operating systems design and implementation OSDI '08*, San Diego, December 2008

Chapter 4
Designing Distributed Services Creation, Service Delivery and Service Assurance with the Architectural Resilience of Cellular Organisms

Abstract The DIME computing model is implemented in two platforms to demonstrate its feasibility and evaluate its usefulness: 1. DIMEs in Linux approach demonstrates the encapsulation of a Linux process as a DIME to demonstrate dynamic reconfiguration of service regulation to implement self-repair, auto-scaling, performance management etc., and 2. A native operating system called Parallax encapsulates each core into a DIME in a many-core server to demonstrate the implementation of a distributed service workflow with dynamic FCAPS management of distributed transactions. This chapter discusses how these prototypes could influence the next generation distributed services creation, delivery, and assurance infrastructure.

The Dial-Tone Metaphor and the Service Creation, Delivery and Assurance Platforms

"La plus que ça change, la plus que c'est la même chose?"

Although, it is not fashionable in the current IT circles, we use the dial-tone metaphor to describe service connection management with telecom grade trust between service providing computing engines and service consuming computing engines. Originally, the dial-tone was introduced to assure the telephone user that the exchange is functioning when the telephone is taken off-hook by breaking the silence (before an operator responded) with an audible tone. Later on, the automated exchanges provided a benchmark for telecom grade trust that assures managed resources on-demand with high availability, performance and security. Today, as soon as the user goes on hook, the network recognizes the profile based on the dialing telephone number. As soon as the dialed party number is dialed,

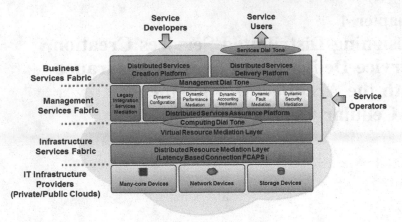

Fig. 4.1 Service creation, delivery and assurance reference model

the network recognizes the destination profile and provisions all the network resources required to make the desired connection, to commence billing, to monitor and to assure the connection availability, performance and security till one of the parties initiates a disconnect. The continuous visibility and control of the connection allows service assurance even in the case of an earthquake or any such natural disaster. Call waiting, call forwarding, 800 service call model and multi-party conferencing—all these features contribute to the "telecom-grade trust" that the telecommunication network has come to symbolize.

The reference model [1] shown in Fig. 4.1, identifies three dial-tones, namely, (1) The *resource dial-tone* that assures computing element resources (CPU, memory, network bandwidth, storage capacity, throughput ad IOPS) on demand, (2) The *service management dial-tone* providing FCAPS management services and signaling services for use in various computing service workflow creation, and delivery, and (3) The *service delivery dial-tone* that delivers and assures the service at run-time to end users who use the services on demand.

The reference model describes the relationships of various stakeholders (1) Infrastructure Providers, (2) Service Providers, (3) Service Developers, and (4) End Users. Below, we revisit how the reference model will affect, benefit and be deployed by each of the stake holders.

Infrastructure providers: These are vendors who provide the underlying computing, network and storage infrastructure that can be carved up into logical clouds of computers which will be dynamically controlled to deliver massively scalable and globally interoperable service network infrastructure. The infra-structure will be used by both service creators who develop the services and also the end users who utilize these services. This is very similar to switching, trans-mission and access equipment vendors in the telecom world who incorporate service enabling features and management interfaces right in their equipment. Current storage and computing server infrastructure has neither the ability to dynamically dial-up and dial-down resources nor the capability for dynamic usage-aware management which will help eliminate the numerous layers of present day management systems contributing to the total cost and human latency involved. The new reference architecture provides requirements for the infrastructure vendors to eliminate current systems administration oriented management paradigm and enable next generation real-time, on-demand, FCAPS-based management so that applications can dynamically request the dial-up and dial-down of allocated resources.

Service providers: With the deployment of the infrastructure satisfying the requirements of the new reference architecture, service providers will be able to assure both service developers and service users that resources will be available on demand. They will be able to effectively measure and meter resource utilization end-to-end usage to enable a dial-tone for computing service while managing service levels to meet the availability, performance and security requirements for each service. The service provider will now manage the application's connection to computing, network and storage resource with appropriate service level agreements. This is different from most current cloud computing solutions that are nothing more than hosted infrastructure or applications accessed over the Internet. This will also enable a new distributed virtual services operating system that provides distributed FCAPS-based resource management on demand.

Service Developers: They will be able to develop cloud-based services using the management services API to configure, monitor and manage service resource allocation, availability, utilization, performance and security of their applications in real-time. Service management and service delivery will now be integrated into application development to allow application developers to be able to specify run time service level agreements.

End Users: Their demand for choice, mobility and interactivity with intuitive user interfaces will continue to grow. The managed resources in the reference architecture will now not only allow the service developers to create and deliver services that end users can dynamically access on devices of their choice, but also enable service providers with the capability to provision in real-time to respond to changing demands, and to charge the end-users by metering exact resource usage for the desired service levels.

DIME Network Architecture and the Resilient Service Creation, Delivery and Assurance Environment

In its essence, the von Neumann computing model implements both service and its regulation in a serial fashion using a Turing machine. Even with extensions to the von-Neumann computing model, such as cache memory, virtual memory and multi-threading, the service and its regulation are specified at compile time, executed serially and management cannot be controlled at run time. Over time, the static nature of service control which originated from the server-centric administrative paradigm is compensated by myriad administrative systems, specialized hardware solutions and cross-domain management systems resulting in the increase of both cost and complexity.

The DIME network architecture on the other hand, exploits the parallelism to address the temporal phenomena involved in assuring transaction integrity in a distributed system. Figure 4.2 shows the comparison between von-Neumann model of service implementation and DIME network based service implementation.

Louise Barrett [2] making a case for the animal and human dependence on their bodies and environment—not just their brains—to behave intelligently, highlights the difference between Turing Machines implemented using von Neumann architecture and biological systems. "Although the computer analogy built on von Neumann architecture has been useful in a number of ways, and there is also no doubt that work in classic artificial intelligence (or, as it is often known, Good Old Fashioned AI: GOFAI) has had its successes, these have been somewhat limited, at least from our perspective here as students of cognitive evolution." She argues that the Turing machines based on algorithmic symbolic manipulation using von Neumann architecture, gravitate toward those aspects of cognition, like natural language, formal reasoning, planning, mathematics and playing chess, in which the processing of abstract symbols in a logical fashion and leaves out other aspects of cognition that deal with producing adoptive behavior in a changeable environment. Unlike the approach where perception, cognition and action are clearly separated, she suggests that the dynamic coupling between various elements of the system, where each change in one element continually influences every other element's direction of change has to be accounted for in any computational model that includes system's sensory and motor functions along with analysis. This emphasis on the sensory monitoring of the environment, dynamic coupling, connectivity and system-wide coordination is also confirmed by observations on cell communication. As mentioned in Chap. 2, according to biologist Sean B. Carroll [3], DNA determines when, where, and how much of a gene's product is made. Animal architecture is a product of genetic regulatory network architecture.

Cellular organisms developed very sophisticated computing models well before their brains evolved. The architectural resiliency of cellular organisms stems from their ability to manage highly temporal phenomena. System-wide connectivity and coordination require a sense of time, history and synchronization between various tasks performed by a group of loosely coupled elements which, as Louis Barrett points out, the Turing machine implemented using the stored program control lacks. Discussing the nature of temporal phenomena, she writes "This means simply that the actual rates and rhythms that characterize a particular process play an important and central role in getting the job done. This could be the way that the underlying physical processes of the brain work (how long it takes for a neurotransmitter, like nitric oxide or glutamate, to diffuse through the brain, for example, or how long it takes for such neurotransmitters to modulate neuronal activity), which in turn could affect the specific duration or rates of change in other

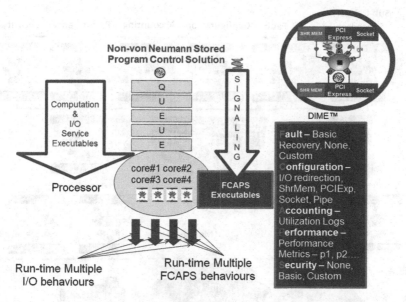

Fig. 4.2 The separation of service and its regulation using parallelism, signaling and self-management abstractions. A video explains the non-von Neumann behavior with parallel signaling overlay over the serial von-Neumann computing network (http://youtu.be/j13qAPZR6G8)

physiological processes. Similar intrinsic rhythms in the body may also be important, as will other aspects of the body dynamics that relate to, for example, the mechanical properties of the muscle, which dictate where and how fast an animal can move. These bodily processes may, in turn, need to be synchronized precisely with temporal processes occurring outside of the animal in the environment." She also points out that the coordination and synchronization requires system-wide information processing and routing that the brain provides.

Compare this with the quest for real-time information processing currently being driven by global communication, collaboration and commerce at the speed of light. Whether it is high frequency trading, web-based commerce, social networking or federated enterprise computing, the ability to manage highly temporal phenomena in real-time is becoming critical. System-wide connectivity, high availability, security and performance management require coordination with a sense of time, history and synchronization between various tasks performed by a group of loosely coupled elements.

Figure 4.3 shows the DIME network service delivery and assurance infrastructure.

The ability of the DIME network to monitor and control the service through the parallelization of service delivery and its regulation decouples the services management from the underlying hardware infrastructure management. For example, if hardware that supports a particular DIME fails, the fault management policies monitoring the *service heartbeat* will immediately kick-in the

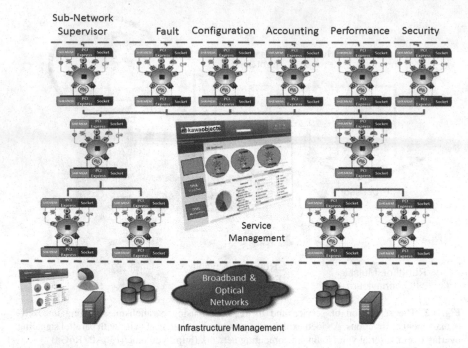

Fig. 4.3 Service delivery and assurance infrastructure showing the decoupling of services management from infrastructure management

recovery policies both at the node level and at the network level. The services deployed either in the DIME node or a sub-network of DIMEs, which are affected by the hardware, are appropriately recovered based on the policies independent of the operating system or the hardware configuration of the hardware host. This is in contrast to current cloud architecture where the services are not independent of the local operating system (in this case a virtual server) and the server configuration.

The decoupling of services management from the underlying hardware infrastructure management allows designing and deploying highly reliable services without requiring highly reliable clusters and specialized enterprise class hardware. The resulting simplification and commoditization of infrastructure hardware hopefully, reduces costs of transactions and improves resiliency of service delivery.

DIMEs in Linux Implementation

Figure 4.4 shows the implementation of DIME network architecture in Linux operating system [4]. Each Linux process is encapsulated in a DIME in which the service regulation and service execution are implemented in parallel. The service regulator defines the service fault, configuration, accounting, performance and

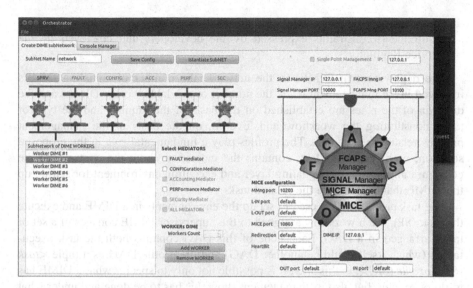

Fig. 4.4 The screenshot of DIMEs in Linux orchestrator creating the DIME network

security policies which are executed as parallel threads control the loading and executing of the service executable.

DIME network with parallel signaling and computing workflows is implemented using two classes of DIMEs:

1. Signaling DIMEs responsible for the management layer at the network level are of the type Supervisor and the Mediator. The Supervisor sets up and controls the functioning of the sub network of DIMEs where the workflow is executed. It coordinates and orchestrates the DIMEs through the use of the Mediators. A Mediator is a specialized DIME for providing predefined roles such as fault or configuration or accounting or performance or security management. The Configuration Manager performs network-level configuration management and provides directory services. These services include registration, indexing, discovery, address management and communication management with other DIME networks. The Fault Manager guarantees the availability and reliability in the sub network by coordinating the "Fault" components of the FM of all the DIMEs involved in the workflow provisioning. The Fault Manager DIME detects and manages the faults in order to assure the correct completion of the workflow. The Performance Manager coordinates performance management at the network level and coordinates the performance using the information received through the signaling channel from each node. The Security Manager assures network level security by coordinating with the individual DIME component security. The Account Manager tracks the utilization of the network wide resources by communicating with the individual DIMEs.

2. Worker DIMEs constituting the "execution" layer of the network perform domain specific tasks that are assigned to them. A worker DIME, in practice,

provides a highly configurable execution environment built on the basis of the requirements/constraints expressed by the developers and conveyed by the Service Regulator.

The deployment of DIMEs in the network, the number of signaling DIMEs involved in the management level, the number of available worker DIMEs and the division of the roles are established on the basis of the number and the type of tasks constituting the workflow and, overall, on the basis of the management profiles related to each task. The profiles play a fundamental role in the proposed solution; each profile, in fact, contains the indication about the control and the configuration of both the signaling layer and execution environment for setting up the DIME that will handle the related task.

The task profile (SR) is used to set up the environments in a DIME and execute the task (SP). Each workflow assigned to the Supervisor DIME consists of a set of tasks arranged in a DAG. Each node of this DAG contains both the task executables (which itself could be another DAG) and the profile DAG as a tuple *<task (SP), profile (SR)>*: in this way, it is possible not only to specify what a DIME has to do or execute but also its management (how this has to be done and under what constraints). These constraints allow the control of FCAPS management both at the node level, the sub-network level, and the network level.

The supervisor DIME, upon receiving the workflow, identifies the number of tasks and their associated profiles. It instantiates other DIMEs based on the information provided, by selecting the resources among the ones available, both the management and the execution layers. In particular, the number of tasks is used to determine the number of needed DIMEs while the information within the profiles becomes instrumental to define (1) the signaling sub-network, (2) the type of relationship between the mediator DIMEs composing the signaling sub-network and the FM of each worker DIME and, finally, (3) the configuration of all the MICEs of each worker DIME to build the most suitable environment for the execution of the workflow. In this way, the Supervisor is able to create a sub-network that implements specific workflows that are FCAPS managed both at management layer (through the mediators) and at execution layer (through the FM of each worker DIME).

Figure 4.5 shows the DIME orchestrator performing the functions of the FCAPS supervisory DIME (creating worker DIMEs and implementing policies during run-time) and two worker DIMEs executing same application (reading a number from a shared data store incrementing it by one unit and storing it) in parallel. The prototype demonstrates following features:

1. DIME worker fault management which assures when a heartbeat fails from a worker DIME, the orchestrator re-instantiates the worker, re-loads and executes the program
2. Dynamic redirection of input/output to a file or another DIME
3. Monitoring and managing performance parameters from each DIME (received periodically by querying the system) and

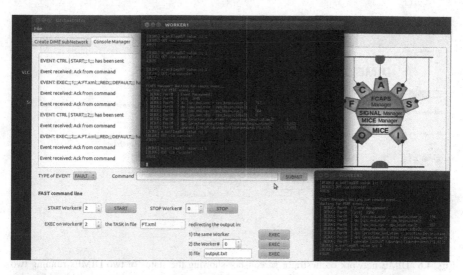

Fig. 4.5 Screenshot showing the execution of "counter.app" by two worker DIMEs that are regulated by the orchestrator DIME implementing FCAPS management

4. Simple security check with login authentication before executing regulation commands by each worker.
5. Network-wide auto-scaling, self-repair, performance monitoring and management and distributed workflow execution

The details of DIMEs in Linux implementation are discussed in [4]. This paper demonstrates the implementation of parallel signaling channel for service management and demonstrates auto-scaling, self-repair and performance management of Linux processes encapsulated as DIMEs.

DIMEs in Multi-Core Server Using a Native Operating System Called Parallax

The Parallax implementation [5, 6] demonstrates the auto-scaling, self-repair and dynamic input/output redirection features supported by DIME network architecture. A DIME network orchestrator is used to instantiate and provide FCAPS management of a DIME network implemented across multiple multi-core servers. Figure 4.6 shows the orchestrator screenshot along with discovery and FCAPS management menus. Figure 4.7 shows self-repair execution. The figure shows the screen for dynamically reconfiguring FCAPS parameters of each application at run time and the application status.

Two programs "Counter.app" with self-repair policy associated with it and the "helloWorld.app" with no self-repair policy associated are shown running before

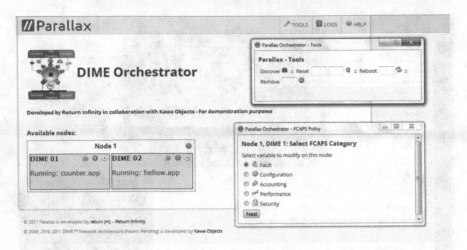

Fig. 4.6 DIME network orchestrator screenshot showing the status of two DIMEs running two different applications running on Parallax operating system in a multi-core server (A video demonstrates the input/output redirection, self-repair and auto-scaling features supported by Parallax http://youtu.be/IMXxmRSVGoI)

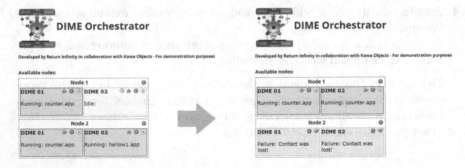

Fig. 4.7 Application (counter.app) running in two DIMEs (before and after recovery)

the hardware fault on the server running these applications. After the hardware fault, the application with self-repair policy is automatically recovered on a new server where a free DIME was available and the program with no recovery policy associated is not recovered. The details of implementation of Parallax are described in [5, 6]. Implemented in assembler with C and C++ interface, Parallax has a very small footprint and encapsulates each core into an FCAPS managed and signaling enabled DIME. An orchestrator allows creating services with service regulation and service executable packages and orchestrates the workflow based on policies. The orchestrator is used to demonstrate auto-scaling, self-repair, and input/output redirection during run-time.

In summary, both DIMEs in Linux and Parallax approaches have demonstrated the feasibility of service management separation from service execution and dynamic reconfiguration of service regulation to implement self-repair, auto-scaling, performance management etc. The purpose of this research brief is to propose a new approach different from conventional computing and current cloud and grid computing approaches and to demonstrate its feasibility. These approaches demonstrate self-repair, auto-scaling and live migration, albeit on a small prototype scale, without the use of Hypervisor or a plethora of management systems. In order to take this research to next level, it requires larger participation from the research community. Only such an effort with an open mind will decide whether this approach has any merit. Given the established and vested interests in existing approaches it is not easy to get attention to new ideas either through academic research or venture capital. This research brief is an open call for such an effort.

References

1. V. Sarathy, P. Narayan, R. Mikkilineni. Next Generation Cloud Computing Architecture—Enabling Real-Time Dynamism for Shared Distributed Physical Infrastructure, *Proceedings of IEEE WETICE2010-Enabling Technologies: Infrastructures for Collaborative Enterprises (WETICE), 19th IEEE International Workshop on 2010*, pp. 48–53
2. L. Barrett, *Beyond the Brain: How Body and Environment Shape Animal and Human Minds* (Princeton University Press, Princeton, 2011), pp. 116, 122
3. S.B. Carroll, *The New Science of Evo Devo—Endless Forms Most Beautiful* (W. W. Norton & Co., New York, 2005), pp. 12, 106, 113, 129
4. G. Morana, R. Mikkilineni, Scaling and Self-Repair of Linux Based Applications Using a Novel Distributed Computing Model Exploiting Parallelism. *Enabling Technologies: Infrastructure for Collaborative Enterprises (WETICE) 2011 20th IEEE International Workshops on*, pp. 98–103, 27–29 June 2011
5. R. Mikkilineni, I. Seyler, Parallax—A New Operating System for Scalable, Distributed, and Parallel Computing. *Parallel and Distributed Processing Workshops and Phd Forum (IPDPSW), 2011 IEEE International Symposium on*, pp. 976–983, 16–20 May 2011
6. R. Mikkilineni, I. Seyler, Parallax—A New Operating System Prototype Demonstrating Service Scaling and Self-Repair in Multi-core Servers. *Enabling Technologies: Infrastructure for Collaborative Enterprises (WETICE) 2011 20th IEEE International Workshops on*, pp. 104–109, 27–29 June 2011

Chapter 5
Dime Network Architecture: Future Research Directions and Conclusion

Abstract The DIME computing model, with FCAPS management and signaling, allows establishing equilibrium patterns and monitor and control exceptions system-wide. It allows contention resolution based on system-wide view and eliminates race conditions and other common issues found in current distributed computing practice. In systems with strong dynamic coupling between various elements of the system, where each change in one element continually influences other element's direction of change, signaling helps implement system-wide coordination and control based on system-wide priorities, workload fluctuations, and latency constraints. These features are used to identify some future research directions. In order to take these concepts to practical application in mission critical environments, the DME network architecture based prototypes require validation and acceptance by a larger community.

Designing a New Class of Distributed Systems Using DIME Network Architecture

"Prediction is not therefore a simple concept, especially when one has the notion of time to incorporate. The nature and complexity of what one extrapolates from, the precision with which the processes of development are thought to be known, whether the outcome predicted has a contaminating effect on the prediction in question and may thus modify, how far into the future this extrapolation is intended to predict, the range of variables which can be accommodated in calculations; all these are some of the many and more obvious problems which make foretelling the future a hazardous business"—Leo Howe, "Predicting the Future", edited by Leo Howe and Alan Wayne, Cambridge University Press, 1993, p. 4.

Current work on DIME network architecture was first presented in WETICE 2010 in Larissa, Greece based on the workshop discussions started in WETICE 2009 in Groningen, The Netherlands.

R. Mikkilineni, *Designing a New Class of Distributed Systems*, SpringerBriefs in Electrical and Computer Engineering, DOI: 10.1007/978-1-4614-1924-2_5, © The Author(s) 2011

The DIME network architecture departs from conventional von Neumann computing model implementing a Turing machine. It adds self-monitoring and self-control of each Turing computing node and a parallel signaling enabled network to implement the management of temporal behavior of workflows executed as directed acyclic graphs using a network of managed Turing machines. The two prototypes demonstrate that the parallel signaling overlay and continuous monitoring and control (at specified interval based on business priorities, workload fluctuations and latency constraints) enable programming auto-scaling, self-repair, performance optimization and end-to-end transaction management. The signaling abstractions uniquely differentiate this approach from conventional computing or the grid and cloud strategies.

Signaling in the DIME network architecture is as important as it is in cellular organisms to provide resilience [1, 2]. In summary, the DIME network architecture adopts the following key abstractions:

1. Parallel signaling channel for monitoring and control of a distributed network of autonomous computing elements (the Turing machines),
2. Programmable self-managing capabilities at the node and the network level providing a way to create a blueprint for the business workflow (managed Turning machine network) and
3. A mechanism to monitor and execute FCAPS policies based on business priorities, workload fluctuations and latency constraints.

The recursive network composition abstractions with network, sub-network, and node level, combined with signaling overlay provide a powerful "network" effect that has been exploited in biology and other domains as we attempted to demonstrate in this research brief. This approach is in contrast to the current approaches [3–10] that use von-Neumann computing model for service management where management and execution of services are serialized both in the node (operating system) and the network (a plethora of resource and service management systems). The demonstration of live migration of services is accomplished by DIME networks depending on end-to-end service level monitoring and control of distributed transactions as opposed to resource management at each node.

The advent of many-core severs with hundreds and even thousands of computing cores with high bandwidth communication among them makes the current generation server, networking and storage equipment and their management systems which have evolved from server-centric and bandwidth limited architectures completely unsuited to use in the next generation computing infrastructure efficiently. It is hard to imagine replicating current TCP/IP based socket communication, "isolate and fix" diagnostic procedures, and the multiple operating systems (that do not have end-to-end visibility or control of business transactions that span across multiple cores, multiple chips, multiple servers and multiple geographies) inside the next generation many-core servers without addressing their shortcomings. In order to cope with the scaling issues and utilize many-core technologies effectively, next generation service architecture has to emulate the architectural resiliency of cellular organisms that tolerate faults and implement command and control structures which enable execution of self-configuring, self-monitoring, self-protecting, self-healing and self-optimizing (in short self-*) business processes. We argue that *the recursive network nature of many-core servers with different bandwidths at different levels is ideally suited to exploit the DIME network architecture*.

The DIME network architecture offers new directions of research to provide next level of scaling, telecom grade trust through end-to-end service FCAPS optimization and reduced complexity in developing, deploying and managing distributed federated software systems executing temporal business workflows. To our knowledge, the suggestion to use signaling overlay to manage a Turing machine is proposed for the first time using the DIME computing model in WETICE 2010 [11]. Similarly, the separation of service execution and its management are implemented in the Parallax operating system for the first time at the operating system level. For example every open(), close(), read() and write() operations are part of dynamically reconfigurable operations made possible by parallel signaling channel. This implementation of signaling in the operating system allows the service execution to be dynamically controlled at run time based on FCAPS policies allowing auto-scaling, self-repair, performance monitoring and control, end-to-end transaction security as the two prototypes we have developed demonstrate.

The beauty of the DIME computing model is that it docs not impact the current implementation of the service workflow using von-Neumann SPC nodes. But by introducing parallel control and management of the service workflow, the DIME network architecture improves the scaling, agility and resilience of existing computational workflows both at the node level and at the network level. The signaling based network level control of a service workflow that spans across multiple nodes allows the end-to-end connection level quality of service management independent of the hardware infrastructure management systems. The only requirement for the hardware infrastructure provider is to assure that the node OS provides the required services for the DIME to load the service regulator and the service execution packages to create and execute the DIME network. The parallax OS is designed to do just that.

The network management of DIME services allows different network configurations and management strategies to be dynamically re-configured such as hierarchical scaling using the network composition of sub-networks or peer–peer management systems or client server computing networks. Each DIME with its autonomy on local resources through FCAPS management and its network awareness through signaling can keep its own history to provide negotiated services to other DIMEs thus enabling a collaborative workflow execution.

Future Research Directions

The two implementations, we have demonstrated provide opportunities to migrate existing services to DIME network architecture (encapsulating each process in a conventional operating system in a DIME) or create a new class of self-* service workflows (using a native operating system tuned for the new generation of many-core servers by encapsulating each core as a DIME). This approach allows legacy applications to make themselves avail the resiliency, efficiency and scaling while new applications to take full advantage of programming self-* management.

We identify just a few possible areas of future research that may prove effective:

1. Implementing DNA in current operating systems, as the DIMEs in Linux [12] approach illustrates, provides an immediate path to enhance efficiency of communication between multiple images deployed in a many-core server without any disruption to existing applications. Current generation operating systems, such as Linux and Windows, can support only few tens of CPUs in a single instance and are inadequate to manage servers that contain hundreds of processors, each with multiple cores. The solutions currently proposed for solving the scalability issue in these systems, i.e. the use of single system image, SSI [13] or the introduction of multiple instances of the OS in a single enclosure with socket connectivity (e.g. [14]), are inefficient. For example two Linux images communicate with each other using socket communication even though they are neighbors in the same enclosure with shared memory and PCIExpress availability. The DIME network architecture fills this operating system gap (defined as the difference between the number of cores available in an enclosure and the number of cores visible to a single image instance of an OS) by dynamically switching the communication behaviors from shared memory or PCIExpress or Socket communication depending on a transaction need.

2. Auto-scaling, performance optimization, end-to-end transaction security and self-repair attributes allow various applications currently running under Linux or Windows to migrate easily to more efficient many-core operating platforms while avoiding a plethora of management systems. Implementing DNA on virtual servers in current cloud computing infrastructure such as Amazon AWS or Microsoft Azure by encapsulating a process in conventional OS allows inter-cloud resiliency, efficiency and scaling. In addition, the service management independence from infrastructure management allows a new level of visibility and control to service delivery in these clouds.

3. Implementing a new OS such as Parallax [15, 16] allows designing a new class of scalable, and self-* distributed systems design transcending physical, geo-graphical and enterprise boundaries with true decoupling between services and the infrastructure that they reside on. The service creation and workflow orchestration platforms can be implemented on current generation development environments whereas the run time services deployment and management can be orchestrated in many-core servers with DNA as demonstrated in the prototype.

4. Signaling and FCAPS management implemented in hardware to design a new class of storage could allow the design of next generation IT hardware infra-structure with Self-* properties and application awareness.

5. As hundreds of cores in a single processor enable thousands of cores in a server, the networking infrastructure and associated management software including routing, switching and firewall management will migrate to the server inside from the data center outside. The DIME network architecture with its connection FCAPS management using signaling control will eliminate the need to replicate current network management infrastructure (e.g., TCP/IP, IP address based firewall management etc.) also inside the server. The routing and switching abstractions will be incorporated in intra-DIME and Inter-DIME communication and signaling infrastructure.

6. The separation of services management from the underlying hardware infrastructure management allows a certain relief from denial of services attacks on the infrastructure. For example, the signaling allows detection of poor response and immediate response in case of an attack on a particular portion of the infrastructure.

Eventually, it is possible to conceive of signaling being incorporated in the many-core processor itself to leverage the DNA in hardware.

Conclusion

We argue that the DIME network architecture is a next step in the evolution of computing models from von-Neumann serial computing to a network-centric parallel non-von Neumann computing model where each Turing machine is managed and signaling enabled. Figure 5.1 shows the evolution of network-centric services delivery which started with voice services (which connected billions of humans to communicate with each other) and evolved to the internet based data services (which connected billions of computing devices to exchange data) to the next evolution of collaborating distributed services infrastructure (connecting trillions of individual service modules to collaborate with each other).

Evolution of living organisms has taught us that the difference between survival and extinction is the information processing ability of the organism to:

1. Discover and encapsulate the sequences of stable patterns that have lower entropy, which allow harmony with the environment providing the necessary resources for its survival,
2. Replicate the sequences so that the information (in the form of best practices) can propagate from the survived to the successor,
3. Execute with precision the sequences to reproduce itself,
4. Monitor itself and its surroundings in real-time, and
5. Utilize the genetic transactions of repair, recombination and re-arrangement to sustain existing patterns that are useful.

The DIME network architecture attempts to implement similar behavior in computing architecture to improve the resiliency, efficiency and scaling of

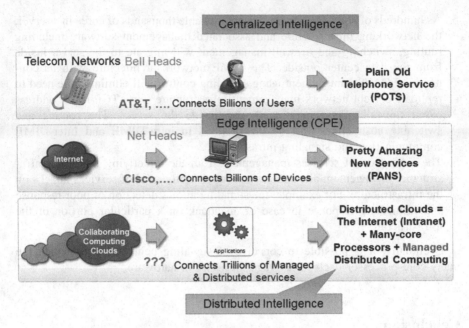

Fig. 5.1 The evolution of network-centric intelligence

computational workflows based on dynamic interactions of various components and their environment. This is made possible by two technology advances—the many-core processors with parallelism and performance required to effectively implement the new computing model and the high bandwidth that allows the temporal dynamics of distributed computing to be effectively managed.

By supporting the four genetic transactions of replication, repair, recombination and reconfiguration, the DIME computing model comes close to what von Neumann was searching for in his Hixon lectures [17]. "The basic principle of dealing with malfunctions in nature is to make their effect as unimportant as possible and to apply correctives, if they are necessary at all, at leisure. In our dealings with artificial automata, on the other hand, we require an immediate diagnosis. Therefore, we are trying to arrange the automata in such a manner that errors will become as conspicuous as possible, and intervention and correction follow immediately." Comparing the computing machines and living organisms, he points out that the computing machines are not as fault tolerant as the living organisms. He goes on to say "It's very likely that on the basis of philosophy that every error has to be caught, explained, and corrected, a system of the complexity of the living organism would not run for a millisecond." The DIME implementation of self-repair using the Parallax operating system and the DIMEs in Linux, described in this research brief, both point to a potential new approach for designing a new class of distributed systems. The purpose of this research brief is to offer an alternative. Only time will tell if the new approach is useful enough to cross the barriers to adoption in mission critical environments.

References

1. M. Mitchell Waldrop, *Complexity: The Emerging Science at the Edge of Order and Chaos* (Simon and Schuster Paperback, New York, 1992), p. 31
2. S.B. Carroll, *The New Science of Evo Devo—endless Forms Most Beautiful* (W. W. Norton & Co, New York, 2005), p. 12, 106, 113, 129
3. D. Patterson, The trouble with multi-core. *Spectrum, IEEE* **47**(7), 28–32, 53 (2010)
4. A. Baumann, P. Barham, P.-E. Dagand, T. Harris, R. Isaacs, S. Peter, T. Roscoe, A. Schupbach, A. Singhania, The multikernel: A new OS architecture for scalable multicore systems. in *Proceedings of the 22nd ACM Symposium on OS Principles*, Big Sky, USA, Oct 2009
5. D. Wentzlaff, A. Agarwal, Factored operating systems (fos): The case for a scalable operating system for multicores. SIGOPS Oper. Syst. Rev. **43**(2), 76–85 (2009)
6. R. Liu, K. Klues, S. Bird, S. Hofmeyr, K. Asanovi'c, J. Kubiatowicz, Tesselation: Space-Time Partitioning in a Manycore Client OS, In HotPar09, Berkeley, 03/2009 (2009)
7. J.A. Colmenares, S. Bird, H. Cook, P. Pearce, D. Zhu, J. Shalf, S. Hofmeyr, K. Asanovic, J. Kubiatowicz, Tesselation: Space-Time Partitioning in a Manycore Client OS. in *Proceedings 2nd USENIX Workshop on Hot Topics in Parallelism (HotPar'10)*. Berkeley, USA, June 2010
8. E.B. Nightingale, O. Hodson, R. McIlroy, C. Hawblitzel, G. Hunt, Helios: Heterogeneous Multiprocessing with Satellite Kernels, ACM, SOSP'09, Big Sky, Montana, USA, 11–14 Oct 2009
9. O. Mao, F. Kaashoek, R. Morris, A. Pesterev, L. Stein, M. Wu, Y. Dai, Y. Zhang, Z. Zhang, Corey: An operating system for many cores. in *Proceedings of the 8th USENIX Symposium on Operating Systems Design and Implementation OSDI '08*. San Diego, California, Dec 2008
10. R. Buyya, C.S. Yeo, S. Venugopal, J. Broberg, I. Brandic, Cloud computing and emerging IT platforms: Vision, hype, and reality for delivering computing as the 5th utility. Future Gener. Comput. Syst. **25**(6), 599–616, ISSN 0167–739X, Elsevier Science, Amsterdam, The Netherlands, June 2009
11. R. Mikkilineni, Is the Network-centric Computing Paradigm for Multi-core, the Next Big Thing? Retrieved July 22, 2010, from Conver-gence of Distributed Clouds, Grids and Their Management: http://computingclouds.wordpress.com
12. G. Morana, R. Mikkilineni, Scaling and Self-repair of Linux Based Services Using a Novel Distributed Computing Model Exploiting Parallelism. *Enabling Technologies: Infrastructure for Collaborative Enterprises (WETICE) 2011 20th IEEE International Workshops on*, pp. 98–103, 27–29 June 2011
13. R. Mikkilineni, I. Seyler, Parallax—A New Operating System for Scalable, Distributed, and Parallel Computing. *Parallel and Distributed Processing Workshops and Phd Forum (IPDPSW), 2011 IEEE International Symposium on*, pp. 976–983, 16–20 May 2011
14. R. Buyya, T. Cortes, H. Jin, Single system image. Int. J. High Perform. Comput. Appl. **15**(2), 124–135 (2001)
15. www.seamicro.com
16. R. Mikkilineni, I. Seyler, Parallax—A New Operating System Prototype Demonstrating Service Scaling and Self-Repair in Multi-core Servers. *Enabling Technologies: Infrastructure for Collaborative Enterprises (WETICE) 2011 20th IEEE International Workshops on*, pp. 104–109, 27–29 June 2011
17. J. von Neumann, Theory of Natural and Artificial Automata, in *Papers of John von Neumann on Computing and Computer Theory*, ed. by W. Aspray, A. Burks (MIT Press, 1987), p. 408, 474. (Charles Babbage Institute Reprint Series for the History of Computing vol. 12.)